T0313994

CODE WORK

PRINCETON STUDIES IN
CULTURE AND TECHNOLOGY

Princeton Studies in Culture and Technology
Tom Boellstorff and Bill Maurer, series editors

This series presents innovative work that extends classic ethnographic methods and questions into areas of pressing interest in technology and economics. It explores the varied ways new technologies combine with older technologies and cultural understandings to shape novel forms of subjectivity, embodiment, knowledge, place, and community. By doing so, the series demonstrates the relevance of anthropological inquiry to emerging forms of digital culture in the broadest sense.

For a full list of titles in the series, go to https://press.princeton.edu/series/princeton-studies-in-culture-and-technology.

Code Work

Hacking Across the US/México Techno-Borderlands

Héctor Beltrán

Illustrations by Daniela Rivero

PRINCETON UNIVERSITY PRESS

PRINCETON AND OXFORD

Published by Princeton University Press
41 William Street, Princeton, New Jersey 08540
99 Banbury Road, Oxford OX2 6JX

press.princeton.edu

ISBN 9780691245034
ISBN (pbk.) 9780691245041
ISBN (e-book) 9780691245058

British Library Cataloging-in-Publication Data is available

Editorial: Fred Appel and James Collier
Production Editorial: Jaden Young
Jacket/Cover Design: Karl Spurzem
Production: Lauren Reese
Publicity: William Pagdatoon
Copyeditor: Catja Pafort

Jacket artwork by Daniela Rivero

This book has been composed in Adobe Text and Gotham

10 9 8 7 6 5 4 3 2 1

to ma and pa: Borderlands hackers par excellence

CONTENTS

ILLUSTRATIONS

by Daniela Rivero

ACKNOWLEDGMENTS

Writing this book has really been an iterative adventure, riding the loops and returns, insides and outsides of the code worlds. I am indebted to the teachers, mentors, friends, students, colleagues, and haters who have guided me, nourished me, and challenged me during the different stages of the ride. I now *return* this book in the hopes that it does justice to some of these code workers and that it might provide some inspiration for others to take off on their own iterations.

The acknowledgments section is my favorite part of a book and the debts are deep in this now decade-long project, so this might be long.

Carlos Briggs was the first person to invite me into the world of academia and to challenge me by giving me the tools to understand how things work across different domains, to help me see that one might be able to hack these worlds by sharpening the tools one is already carrying. There are no words to describe your dedication, relentlessness, and support; it is an honor to be your student, mentee, and friend, Carlos. Thank you to other members of my dissertation committee in the Anthropology Department at Berkeley. Aihwa Ong, for daring me to make my writing clearer and my interventions bolder. Daniel Fisher, for your encouragement and for your careful engagement with my ideas.

My other key mentor at Berkeley was Patricia Baquedano-López: thank you for your sincere companionship in the academic worlds and for reminding me to always keep present our larger purpose as teacher-mentors. Patricia gave me the confidence to run with different projects at the Center for Latino Policy Research, where many of the ideas on tech work and diversity in this book had their origins as part of the Latinxs and Tech Initiative. Several undergraduate students worked with me to organize the "Diversity in Tech: Beyond Bootstraps" conference: Héctor "Tito" Callejas, Diana Arteaga, Yesenia Luis, Ulises Serrano, and Chelsea Burroughs. At that special conference, we were able to foment conversations and relationships that have stood the test of time and iterations of all sorts. From the academic world: Dolores Inés Casillas, Omar Ruvalcaba, Lucy Carrillo, and Blanca Gordo; and from the tech industry, nonprofit, and startup world: Orson Aguilar, Karla Monterroso, Jacob Martinez, Nicole Sanchez, Laura Gómez, Andrea Guendelman, Sarahi Espinoza, the homey Eutiquio "Tiq" Chapa, and the one and only, the real "Abogadazo," Diego LaFuente.

Three key teacher-mentors from the Ethnic Studies Department taught me about epistemological struggle within the university. Taking classes with Ramón Grosfoguel and listening to his theories that thought against *their* theories was medicinal; Lok Siu's course on ethnography further taught me to challenge ethnographic methodologies; Ray Telles' humility and patience teaching undergraduates helped to keep me grounded.

The Institute for the Study of Societal Issues (ISSI) was fundamental to my intellectual development. Deborah Freedman Lustig was always gracious enough to read drafts of works in progress, and her feedback was incorporated into Chapter 1 before it was an article or chapter. David Minkus kept us all on our toes with absurdly real connections across seemingly unrelated domains. My fellow cohort members reminded me to resist disciplinary boundary-work. Thank you, Deborah and David, and Esther Yoona Cho, Jen Rose Smith, Cynthia Ledesma, Zawadi Rucks-Ahidiana, Melody Tulier, and Kelechi Uwaezuoke. Keith Feldman provided invaluable feedback in his role as respondent during my ISSI symposium presentation. Constructive criticism from Clara Mantini-Briggs during this presentation also proved fruitful; and her consejos and acompañamientos across the years were vital.

My student colleagues across cohorts, departments, and institutions added the necessary laughs and drama to remain somewhat sane during my graduate school tenure. Shout out to Lisa Min, Antonia Rivas, Samuele Collu, Danae Valenzuela Aramburo, Felipe González, Andrés Jacobo García Molina, Kevin Kenjar, Rosa Norton, Kim Tran, Kristin Sangren, Mila Djordjevic, Alexa Hagerty, Blanca Gamez-Djokic, Krista Cortes, Teresa Montoya, Daniel Woo, Paulino Malcolm, Vreni "Chhoti Maa" Michelini, Chantiri Resendiz, Alejandro Jimenez, Tomás Monarrez, Emilia Tortuga, Camilo Lund-Montaño—and anybody I am forgetting who passed through Casa Azul.

My graduate studies were supported by generous support from the Ford Foundation Predoctoral Fellowship. My dissertation research was supported by the UC-Mexus Dissertation Research Grant, the Wenner-Gren Foundation for Anthropological Research Dissertation Fieldwork Grant, and the Social Science Research Council International Dissertation Research Fellowship. Preliminary fieldwork was supported by the Center for Race and

Gender and the Center for Latin American Studies at UC Berkeley. Dissertation writing was supported by the Mellon/ACLS Dissertation Completion Fellowship and by the School for Advanced Research (SAR) Resident Scholars Program.

My year at SAR might have been one of the best years of my life. What a luxury to be able to disconnect and rethink the dissertation content under New Mexico sunsets. I am indebted to the staff at SAR, and especially to Paul Ryer and Maria Spray, for making my family's stay comfortable and for providing the ideal setting for writing. Fellow resident scholars provided important feedback on early versions of Chapter 1 and Chapter 5. Thank you, Milena Melo, Brian Smithson, and Deanna Dartt, and especially Pierrette Hondagneu-Sotelo, who was not only my office neighbor but helped me see that there are different ways to be an academic. The SAR Anne Ray interns, who were ideal neighbors and confidantes, make me feel nostalgic about those Santa Fe days; thank you, Brenna Two Bears and Samuel Villarreal Catanach.

As I was finishing up graduate school, I presented an early version of Chapter 1 as part of the "Fabricating Utopics" panel I co-organized with Erin Mariel McElroy at the 2017 Annual Meeting of the American Anthropological Association. Erin, along with fellow panelists Joshua Zane Weiss and Caroline Kao, provided critical commentary, and our brilliant moderators Sareeta Amrute and Luis Felipe Murillo not only guided our proposed interventions but took our panel to new levels with a special issue for *Catalyst: Feminism, Theory, Technoscience* titled "Computing in/from the South." Sareeta and Luis Felipe's theoretical formulations in the introduction to that issue provided key concepts to launch my thinking on "hacking from the South," and I listen closely to anything they have to say about computing otherwise.

Along this same vein of rethinking computing from below, three more scholars have been key interlocutors. Biella Coleman's ethnography with open-source hackers is my canonical text in the anthropology of hacking and a book I always return to in order to remind myself of the promise of anthropology. Lilly Irani's writing on design and innovation from the South was an early motivator for choosing hackerspaces as my research sites. An interview with Christina Dunbar-Hester for the CaMP Anthropology blog (organized by Ilana Gershon) on her *Hacking Diversity* book served as a critical conversation to think deeply about the politics of scaling, othering, visibility, and intersections as they related to hacking cultures. Thank you all for your intellectual mentorship and friendship.

Rihan Yeh inspired this book from its infancy, when she graciously agreed to carefully read through chunks of my dissertation that ended up assembled across different chapters of the book. I fell in love with Rihan's masterful writing about the border and its physical and metaphorical crossings; it was an honor to interview her on her book. Rihan's graduate students at El Colegio de Michoacán further provided insightful comments on an early version of Chapter 1. I hope this final version does not do you too great a disservice. Gracias, Rihan!

I've never been to the University of Chicago, but as I map out my thank yous, I realize that many of the scholars that provided feedback and that have mentored me emanate from that institution, so I think I have to thank the University of Chicago as well?

But let me back up for a second. I learned of Rihan's work through Jonathan Rosa, whose provocations stemming from the intersection of Latinx studies and anthropology have been an inspiration from the very beginning, when I read his PhD dissertation for an ethnography class. Despite Jonathan's prolific publishing in all genres (most notably social media) he has never been too busy to check in. Arlene Dávila's influential writing on Latinidad, its spins and its permutations, has likewise animated my scholarship. Arlene in particular opened up her doors to me when I was a very young and impressionable graduate student. My deepest gratitude to you both for your continued mentorship and unrelenting advocacy.

During the "Digitizing Race—Making Latinxs in the 21st Century," a conference organized by Arlene's "The Latinx Project" at NYU, I not only got to meet and present along with the mighty Ruha Benjamin, but met with one of my future co-conspirators across the techno-Borderlands, Iván Chaar López. What an honor—thank you, Arlene and all of the organizers.

Other friends and colleagues who cut across these events and categories throughout the years, who have motivated me with their work and their good spirits: Adrián Felix, who was one of the first people to make me consider graduate school (this is all on you, compa); Juan Herrera, whose real conversations helped me survive Berkeley; and Jessica López-Espino, who bravely continued to travel the anthro worlds with me after a chance encounter in Kroeber Hall. And of course, the Ford homies: Teresa Montoya, Carlos Jimenez, and Amanda Marin-Collom. From Mexico, I thank Fidele Vlavo for the Montreal and CDMX strolls that resulted in a never-ending flurry of ideas that sometimes went somewhere.

I benefited from a postdoctoral year in the Department of Anthropology at the University of California Irvine, where I was able to initiate the dissertation to book manuscript metamorphosis, and share works in progress as part of a colloquium. I especially thank my mentor Leo Chavez, as well as other faculty members in the department—in particular Bill Maurer, Tom Boellstorff, Eleana Kim, Kim Fortun—as well as Long Bui in the Global Studies department.

Other spaces where parts of my book have been presented and benefited from comments and interventions: Harvard STS Circle, by invitation from Sheila Jasanoff; STS Methods Lab at TIK Centre, University of Oslo, by invitation from Ana Delgado; University College London Department of Anthropology, by invitation from Antonia Walford and Lydia Gibson; Center for Latinx Digital Media at Northwestern University, by invitation from Pablo Boczkowski and Mora Matassi; Centro Multimedia del Centro Nacional de las Artes en México, by invitation from Luis Romero and Leonardo Aranda; UCLA Department of Anthropology, by invitation from Jason Throop and Samy Alim; UCSD Critical Anthropology Workshop, by invitation from Hanna Garth; University of Michigan Department of Anthropology, by invitation and warm welcome from Andrew Shryok, and special hosting from Jason De León; Yale Anthropology Mixed Realities workshop, by invitation from Lisa Messeri and Spencer Kaplan; SSRC Just Tech Platform, by invitation from Catalina Vallejo and Michael Miller; II Encuentro Latinoamericano de Antropología Digital in Chile, by invitation from Natalia Orrego and Iván Flores; and the best for last, Data and Society Research Institute, where chicanocyborg Rigoberto Lara and Ranjit Singh have consistently pushed out the most fun and creative work from scholars in/from the Majority World. A sincere thank you to all for the dialogue and for hosting me!

My MIT family has provided the critical home base from which to sharpen this book's arguments and actually finish it. I have been fortunate to share intellectual and physical space with my wonderful past and present departmental colleagues: Manduhai Buyandelger, Amah Edoh, Michael Fischer, Jean Jackson, Amy Moran-Thomas, Susan Silbey, Bettina Stoetzer, and Chris Walley; as well as graduate or postdoctoral colleagues who are now rising star professors elsewhere—Beth Semel, Luisa Reis-Castro, and Chamee Yang. Our administrative dream team, Carolyn Carlson, Kate Gormley, Irene Hartford, Barbara Keller, and Amberly Steward, have kept our department running like a well-oiled machine. Our inimitable department leaders, Heather Paxson and Stefan Helmreich, provided critical feedback during a book workshop they hosted, and tag-teamed to help me polish later versions of my introduction. Thanks for all you do, Heather and Stefan! My mentor and colleague Graham Jones has been an invaluable interlocutor since before I arrived at MIT, and during the workshop helped me understand that simple conceptual elegance would bring magic to the ethnography, convincing me to resist the urge to pull out my "cheap tricks." Rounding out this all-star cast that made it to my book workshop, Dwai Banerjee further emphasized the necessity of sticking to the ethnography, reminding me that what's there is what's there—and that's just fine.

Outside of my department, Tanalís Padilla has been a phenomenal co-instructor and co-conspirator, and along with Bob and Camilo, also an amazing host to bring some sunshine to the northeast, even when there is none. Likewise, I've been lucky to have Eden Medina as a colleague-mentor and friend, who, along with Cristian and Gael, has provided warmth and joy to my family with their buenas vibras.

Rounding out the MIT characters who deserve a shout out, in these early assistant professor years I've had the privilege to work with and share ideas with brilliant graduate students: Mariel García-Montes, Raha Peyravi, Timothy Loh, Andrea Kim, Ámbar Reyes, Di Wu, as well as Zach La Rock and Hina Walajahi, whose invitation to read on contemporary ethnographic theory and its connections with "writing from below" was a welcome intellectual journey during my last stretch of writing. Not at MIT, but might as well be, check-ins with my Berkeley folklore+anthro comrade Cameron Johnson were always welcome and productive intermissions from the last stages of writing. I'm blessed to continue to work with the next generation of graduate students who will push hacking, making, computing, and all their creative intersections to their limits: Veronica Uribe-del-Aguila, Nicole Hernandez, Jolen Martinez, Lilliana Gil . . .

At Princeton University Press, Fred Appel was an avid supporter of the project from day one. The entire team was tremendous, especially James Collier and Jaden Young, who elegantly marshalled elements of the manuscript through the publication pipeline, and Catja Pafort, who provided pithy suggestions while allowing my techno-Borderlands writing habitus to run wild. The folks at Ideas on Fire were sensational, especially Jean Lee Cole, whose razor-sharp insights helped me tighten up lingering connections and gave me the motivation for the last leg of revisions needed to finish this book.

Finally, I've saved perhaps the two most important categories of folks I want to include here for the end. First, the hackers, hacker-entrepreneurs, *casi hackers*, and evaders of these categories, whose moves across the techno-Borderlands inspired me to write this book in the first place. It's been a labor of love, but your stories, experiences, and tenacity motivated me to keep going, and I hope you will recognize some of your work, engage with my theorizing, and, of course, continue to think with me, but most importantly, continue the code work. Some names (peripheral enough to the research) I can mention: The instructors who helped organize the global startup labs at UNAM and in Xalapa: Miguel Amaya, Hiroshi Mendoza, David Flores, Leonel Mendez, Diego Mendez Canon, Amin Manna, Zachary Hulcher, Eusebio "Chevo" Pérez, and Ricardo Crespo Smith. The *casi hackers* and closely coupled friends helped me think with "hacking from the South": Agustín Basañez, Bawixtabay Torres, Christopher Grajales, Julio Álvarez, Karla Camarillo, Leonor Téllez, Rodrigo "Sr. Gonzalez" Gonzalez, Rodolfo Sousa, Diandra Cruz, and Abel Zavala.

And of course, home is where the heart is. Gracias, ma, pa, and my eternally multitasking sis Ivette, for sticking with the graduate school thing, and having faith that someday I would no longer be a student. (I still am though.) Les quiero mucho.

I'm lucky to have four *chaparritas* at home who have accompanied me through most stages of this book. Thank you, Tuna and Flinka, for always being eager to go on the early morning walks, where some of the concepts finally clicked; Daniela, for all your love and realness, and for listening to "ese plática que siempre doy" just one more time; and our latest addition, Omara, whose light keeps shining brighter every day. Las tqm <3

———

An early version of Chapter 1 appeared as a working paper titled "Staging the Hackathon: Codeworlds and Code Work in México" for UC Berkeley's ISSI (Institute for the Study of Societal Issues). An article version of Chapter 1 was published in *American Anthropologist* as "Code Work: Thinking with the System in México." A version of Chapter 4 was published in *Catalyst: Feminism, Theory, and Technoscience* as "The First Latina Hackathon: Recoding Infrastructures from México."

Ethnographic quotes are verbatim unless otherwise indicated. If original quote was in Spanish, I translate to English and include the quote in both languages. In shorter quotes and where the translation seemed especially straightforward, I include only the English translation and add a footnote indicating that the original quote was in Spanish. I also indicate when a quote is in verbatim English mixed with Spanish, or Spanglish, to avoid confusion. I have substituted pseudonyms for all names except for public or government figures, and unless otherwise noted. Some research participants, for example, specifically asked to use their real names to counter invisibilization of their advocacy efforts.

CODE WORK

[0] Introduction

[0] *Todos con el mismo chip*

In the year 2014, three different collectives of promising young people in Mexico boarded buses. All three shared a general disillusionment with Mexican political and economic reforms, and all three had visions about how to leverage new technologies to disrupt business as usual in Mexico and effect meaningful change. They had different destinations, yet they traveled similar roads—but the three buses and the talented young people aboard never intersected or met. Their journeys also met dramatically different ends. One group received prizes and national acclaim. Another made it to Silicon Valley, where they brushed shoulders with renowned investors and entrepreneurs. Most of the young people aboard the third bus wound up dead or disappeared by state authorities.

The first collective boarded *la combi de la ciencia* (the science VW bus/van), a 1992 baby blue Volkswagen Transporter, with a plan to make science accessible to marginalized rural communities. The project was spearheaded by Cristóbal Miguel García Jaimes, who had become known in Mexico as *El Chico Partículas* (The Particle Boy). At the age of 17, he had worked on building the world's cheapest scale model of a particle accelerator. His vision was to bring this miniature on tour using the combi de la ciencia, showcasing it to his paisanos across rural regions in his native state of Guerrero, driving along a route he called "la ruta de la ciencia" (the science route). One of the key projects they carried out with the combi de la ciencia was called *pepenando computadoras* (trash-picking computers), in which they repaired unused computers from Mexico City and donated them to young kids along the science route who didn't own personal computers. "Qué tal si el próximo genio de la computación está en la sierra!" (Perhaps the next computer genius is somewhere in the mountains of Mexico!), he emphatically told reporters. In the many videos across social media and news sites that documented El Chico Partículas and la combi de la ciencia, Cristóbal wore his "Work Hard Dream Big" t-shirt and frequently pointed out that one could easily find "talent" in marginalized rural towns and that with enough will and educa-tion anybody from these communities could "salir adelante" (get ahead in life). Media reports loved to highlight his humble origins and sense of over-coming, and frequently included his exuberant quotes, such as "Si Pedrito no puede bajar de la montaña, los aceleradores de partículas y la combi de la ciencia subirán hasta Pedrito." (If Pedrito can't come down from the mountain, the particle accelerators and the combi de la ciencia will come up to find Pedrito.)

Around the same time as El Chico Partículas and friends were *pepenando* computers between Mexico City and rural Guerrero aboard their mobile science lab, a different group of technology tinkerers boarded a bus headed out of Mexico City, toward the US. Traveling on this mobile hackathon, this "startup bus," they were tasked with representing Mexico in a festival to be held in Austin, Texas. Chosen from and tried among Mexico's best technol-ogy hackers after being active participants in Mexico's hackathon scene, the young people aboard, like those traveling in la combi de la ciencia, also felt that they had to take matters into their own hands in order to imple-ment needed change in Mexico. One of the young men on this startup bus was Javo, who straddled the line between hacker and entrepreneur, having previously worked on technological projects to prevent voting fraud, as well as other more commercial apps that he had pitched to Mexican investors. Disillusioned with the lack of support he had received from state and private entities for any of his proposed projects, the startup bus allowed him to pitch his ideas to a broader audience. But it also gave him the space to conceive a new idea altogether. Experiencing intermittent internet connection on the bus and having just lived through several earthquakes in Mexico City that left residents disconnected and unable to communicate with one another, Javo and his co-founders came up with an app, Pingafy, that would allow people to communicate with each another without needing an internet con-nection. They not only won that hackathon competition, but after years of perfecting their technology and pivoting their ideas, they received millions of US dollars from Silicon Valley investors who saw their project as a promis-ing "disruptive" technology, and their platform was eventually used during protests and natural disasters across the world.

The third collective that boarded a bus also set out to disrupt business as usual. On September 26, 2014, a group of students from the teacher-training college of Ayotzinapa in the state of Guerrero mobilized to attend the annual commemoration of the 1968 massacre, in which the Mexican army killed hundreds of student demonstrators. To arrive at the event in Mexico City, they commandeered several buses, a practice that was quite common for students from the country's rural boarding schools. These schools—called *normales*—were created in the 1920s for children of campesinos (working-class farmers), and the students had developed a reputation for political militancy.[1] The students wore their penchant for disruption proudly, and taking over private buses as transportation to attend a demonstration was a way to display the tactics they had to resort to since the government had abandoned them in lieu of supporting urban, more "modern" citizens. The *normalistas* had evolved over the century, though they still carefully cultivated a culture that privileged collective action, education for the poor, and student voices in all political matters.[2] As the students set out from the city of Iguala that night, they were met with an armed operation that left most of them dead and 43 students disappeared. Although the case is still officially "unresolved," the disappeared students are still "missing" and many accounts link federal forces to the incident, charging that the Mexican Army was responsible for kidnapping and murdering the students, with many of them last seen being dragged off by federal and state authorities.[3]

While these three collectives experienced drastically different fates, their stories nonetheless share key elements and their political aims overlapped. Like El Chico Partículas, the future teachers at the escuela normal in Ayotzinapa imagined a brighter future for young people in Guerrero. One of the requirements to enter this rural school was to come from a family of *campesinos* or from a poor family, and they worked directly with a younger generation to implement self-sustaining projects, often using new agricultural technologies, aimed at achieving a more egalitarian future. The founders of Pingafy even aligned themselves with the mission of educating the poor and working with rural communities. Javo asserted that his team would have been as happy connecting a few students with no internet access in rural Mexico as helping to connect thousands of protesters in Hong Kong. The interventions of the three groups were inflected by class privileges that granted social and physical mobility, but they also enacted a socially conscious "disruptive" spirit; fed up with promises that never materialized, they

1. Padilla (2013) shows that "escuelas normales" are a direct product of Mexico's 1910–1920 revolution and gained prominence in the 1930s as part of a push for land rights and worker's education and political consciousness. Known for their political radicalism, they garnered frequent intelligence reports from Mexico's *Secretaría de Gobernación* (Ministry of the Interior).

2. Padilla, 2022: 2.

3. The mass murders caused national scandal, protests, and political commentary. For a collection of texts from Mexican and other Latin American scholars addressing the incident, and criminalization of particular youth across Latin America more broadly, see Valenzuela, 2015.

took matters into their own hands. However, instead of receiving the awards and acclaim that El Chico Partículas and friends received, or the millions that Javo and his app's co-founders received, the Ayotzinapa students were met with gunshots.

This juxtaposition of initiatives might be less jarring when contextualized in relation to the sociotechnical imaginaries that have historically animated tensions between the old and the new, the traditional and the modern, the indigenous and the mestizo, the urban and the rural, the technical and the non-technical. In contemporary Mexico the celebration of "modern" engineers, scientists, and entrepreneurs—El Chico Partículas and especially Javo, in this case—became part of the nation-making project to stage the potential of technology to fulfill the promise of progress. The promise of *entrepreneurial* engineers and scientists helped to promote a political agenda where young people were asked to appropriate neoliberal discourses about taking initiative, being self-satisfied, not waiting for government, and being "socially conscious."[4] These entrepreneurial engineers were constructed in opposition to the more rural, less "modern," not-technical-enough normalistas, whose projects and initiatives were considered more threatening and a nuisance to the state's path to progress—their "disruptions" were not considered of the respectable variety that El Chico Partículas', and especially Javo's Silicon Valley–focused efforts, came to represent.

Anxieties about Mexico's future were manifested in state economic reforms as well as in myriad initiatives undertaken by young people themselves. El Chico Partículas became a poster boy for the government's Mexico Conectado project, which promised to end the digital divide across the country by "connecting the disconnected." The project established *nodos* (nodes), or *puntos* (points), across the country where young people could gain access to the promised technological infrastructure that would help them get ahead in life. Many of my research participants took part in the first cohort of the original nodo in the state of Veracruz, and I followed their trajectories as they were commissioned to implement nodes in other parts of Mexico. Within these spaces, the government dedicated resources to hackathons that became part of *retos* (challenges) named *Todos con el mismo chip* (Everybody with the same chip). These challenges occurred

---

4. I use "neoliberal" as a logic of governing for optimal outcomes (Ong, 2006), an approach that moves away from all-encompassing Neoliberalism "whole" (Collier, 2009). As scholars have shown, elements usually associated with "neoliberalism" (e.g., efficiency, transparency, forms of enterprising subjectivity) can take unexpected forms on the ground (DeHart, 2010; Hoffman, 2010).

within the nodos, and they aimed to recruit young people who wanted to propose technological solutions to "transform communities and positively impact Mexican lives," according to their promotional materials. They also made clear that computers as well as a particular coding logic and code model of cognition were central to alleviating Mexico's social and economic ills. "Inglés y computación para todos" (English and computation for all) was a slogan consistently featured on promotional materials for the nodes, presented as a viable solution for helping the nation "catch up."

My opening stories of young people boarding buses, mobilizing across different locations and spaces, mean to highlight this feeling of catching up, this neoliberal hustle, that motivated different groups of young people. Disenchanted with the state-sponsored initiatives meant to boost a failing economy, young idealists were nonetheless willing participants in these endeavors. They attempted to sustain a critical stance on these projects even as they were also figuring out how to succeed in the global information economy workforce, and most importantly, how to stay alive within repressive government structures.

Code Work highlights the ways young people position themselves in relation to narratives that promote the promise of technology and modern infrastructures.[5] As government-sponsored nodes were constructed here but not there, as particular forms of disruption and technological entrepreneurship were valued for some people but not for others, young people learned to function inside of a neoliberal economy by using resources at hand and by appropriating the discourses of flexibility and self-management across highly uneven terrain. I focus in particular on young people who are drawn to the world of computer programming, who find value in honing their code work within hackerspaces. Throughout the book, I show how participants cultivate careful hacker ethics to develop a strong sense of self and to use the practice and language of "coding" to negotiate their sociopolitical reality.

From villages in rural Mexico, through urban labs and hackathons, to the iconic reference point of Silicon Valley, these hackers work on exposing the tensions between the socially transformative potential of hacking and its imbrication in the transnational political economy of tech work divided and structured by the US/México border. Mexican and Latinx hackers immerse themselves in the code worlds, develop hacker ethics, and re-interpret coding logics to think with the underlying political, social, or economic system in an attempt to reorganize their relationships with entities who produce

---

5. Anand et al., 2018; Shankar, 2008.

value from their hacking. Across the book's chapters I take the labor of coding as central to my interlocutor's constructions of self. Connecting such self-making idioms of coding to the logics and metaphors programmers deploy across the "stack"—those interdependent layers of hardware components and software protocols that make high-level computation possible—I develop the concept of the *ethno-stack*. The ethno-stack refers to those interleaved personal, interpersonal, sociopolitical, and sociotechnical elements that come together when actors seek to make computation work for a people, an ethnos, here, often a people identified as Mexican or Latinx. *Code work* is the labor my research participants engage in as they create and think with instances of the ethno-stack. As I will explicate, this work unfolds across a US–México network, or the *techno-borderlands*. In this zone, which I myself traveled as an ethnographer, US/Mexican hackers deploy the language of coding to navigate boundaries of nation, race, ethnicity, class, and gender—boundaries, often hierarchical, that coding promises to reconfigure, but that also remain essential to the transnational economy of tech and, thus, also, often resurgent at every step.

[1] Hackathons, Hacker-Entrepreneurs, and Hacker Ethics

Although the collectives aboard the three buses never intersected, one space that *could* have offered an opportunity for them to meet and collaborate was the hackathon. Hackathons take on very different modes of participation depending on the group of organizers and sponsors, but the basic idea of the form is that an interdisciplinary group of (mostly) young people meet and network with other hackers or entrepreneurs over the course of a weekend.[6] They work collaboratively to prototype project ideas that might resolve an issue related to an organizing theme for the event. Themes such as healthcare, economic or social inequality, climate change, and immigration make frequent appearances. Participant teams then present "pitches" to a panel of experts who judge the viability of the proposed projects, which usually offer technological solutions to the identified problem. While thousands of prototypes might be started at these events, many hackathons end similarly,

---

6. To get a sense of how popular hackathons were when I conducted research, an organization dedicated to enumerating the events and their artifacts reported that in 2016 there were at least 3,450 hackathons organized, 200,000 people participating in them, and 13,000 prototypes built in over 100 countries (Quenardel, 2017).

with participants just shaking hands and saying goodbye, and much of what "gets built" never getting built at all.[7]

An existing and vital literature on the social science and culture of hackathons centers around three themes. First, there is a conversation about connections with state-based hacking.[8] Second, many scholars have tracked the re-emergence of the maker movement and a do-it-yourself ethic that defines hacking as broad clever practice.[9] Third, there is a discussion of how hackathons may operate as a training ground for the frequently unpaid grunt work of the knowledge economy.[10]

In the state-based hacking domain, hackathons are advanced by governments with interests in creating both temporary and permanent spaces as models for a new society, where openness, acceptance, discussion, and participation flourish, and where digital technologies can become tools for empowerment. From the state's perspective, the construction of modern hacker and maker spaces represents a newly forming "innovative culture"; the spaces function as efficient and scalable means to "develop," "modernize," and appear economically competitive.[11] Within these spaces young people are encouraged to take matters into their own hands and assume their role as technical, "entrepreneurial citizens."[12] This connection between entrepreneurial subject-making and neoliberal nation-making is not unique to Mexico. Describing projects across Asia and Africa that present entrepreneurs as drivers of forward-thinking, large-scale social change, Lilly Irani observes, "These projects cast entrepreneurs as collaborative rather than agnostic, technical rather than political, and constructive rather than complaining."[13]

The second thread of research emphasizes the crossover between hacking based on manipulating software and hacking based on manipulating materials. With the rise of hackerspaces, we also see a rise in "fab labs" and

7. Irani, 2015: 804. McIntosh and Hardin (2021) conducted an empirical study of nearly 12,000 project code repositories related to the popular Major League Hacking events between 2018 and 2010 to conclude that very few show patterns of consistent development, with only 7% of projects showing any activity after 6 months.

8. Beltrán, 2020a; Irani, 2019; Lindtner, 2020.

9. Jordan, 2017. Maxigas (2012) explores how these movements and corresponding ethics have been characterized by a turn toward the physical, especially spurred by the emergence of new technologies like 3-D printing.

10. Gregg, 2015; Zukin and Papadantonakis, 2017.

11. Beltrán, 2020a.

12. Irani, 2015, 2019.

13. Irani, 2015: 803.

other makerspaces that center "making" as an activity.[14] Here, making is framed as a solution to social and economic struggles by enabling a return to an authentic, deep, and hands-on engagement with the world—one imperiled by the outsourcing and automation of manufacturing and advances in information technology.[15] Hacking and making are thus framed in opposition to passive consumer culture; "prosumers" are now technology producers and engaged citizens who address societal concerns in a hands-on manner and gain the skills to intervene in the market economy, or at least become employable in it.[16]

The third line of research overlaps with the second by emphasizing how Silicon Valley always looms large in the background of the hackathon. Quite literally, many times the representative images and inspirational quotes from famous Silicon Valley entrepreneurs are painted on the walls of hackerspaces.[17] Here, the hackathon event is analyzed as a microcosm of Silicon Valley dynamics, where participants perform mercurial allegiances and work in focused, high-innovation cycles meant to mimic free-market business processes.[18] While Silicon Valley is a diverse region by no means governed by a single ethos,[19] the techno-entrepreneurialism and high-tech capitalism it has come to represent leads to pre-packaged toolkits and guides which have created a type of branding that attempts to emulate professional identities and economic success in innovation hubs around the world.[20]

Across the many hackathons in both Mexico and the US that constituted my research sites, the cultural practices of the tech companies and collectives that make up Silicon Valley were inextricable from the sociotechnical infrastructures my research participants navigated. Many of my research participants aspired to gain employment with prestigious tech companies, and even if they didn't, they ended up doing work for them on a contractual

14. Gauntlet, 2011.

15. Lindtner, 2020; Uribe, 2021.

16. Lombana-Bermudez et al., 2020.

17. Beltrán, 2020b.

18. Jones et al., 2015.

19. Scholars have shown that "Silicon Valley" is home to a range of distinct values and ideologies, from conventional engineering commitments (English-Lueck, 2002), to "laid back" California attitudes (Saxenian, 1996), to narratives of rapid class mobility for specific ethnic groups (Shankar, 2008), to new age philosophies (Zandbergen, 2010), to neoliberal orientations (Marwick, 2013), to countercultural practices (Turner, 2006).

20. Avle et al., 2017. One Silicon Valley–inspired tech startup approach frequently heard in the hackathon circuit, for example, is looking for the "ideal" team of a hacker (a person with programming skills), a hustler (a person with business skills), and a hipster (a person with marketing/design skills).

basis or used their technologies. Silicon Valley is frequently championed as a model for technological innovation, a place where high revenue generation and disruptive technologies are attributed to a culture of competitive collaboration, lean methodologies, and colorblind meritocracy; these cultural practices are said to "level the playing field." At the same time—and especially after major tech companies released demographic data about their workforce—it is critiqued for underlying structures that promote patriarchy, racialization, and exploitation.[21]

The popularity and proliferation of the hackathon in the early 21st century thus provides an entryway for examining the way some hacking formats attempt to encompass both liberatory and market logics. The highly ephemeral but also highly public nature of the hackathon further allows a precise view of how different communities crystallize and evaporate as they align with hacker and entrepreneurial identities. Companies capitalize on hacking energy "from below" to promote their products and have developers work on their technological platforms and infrastructures.[22] At the same time, anthropologists and other researchers have benefited from the space and time compression of the event, which allows them to treat it as an ideal ethnographic site to focus on the imaginative and communicative labor, in addition to the technical work, that enthusiastic participants come to perform within these spaces.[23] What is being "made" at these events is really a set of dispositions and attitudes about how to develop relationships to new technologies, as well as to one another, as "maker," "hacker," or "entrepreneurial" subjects, many times in the face of precarity and marginalization.[24]

What the compressed time of the hackathon offers for many participants is the promise of self-guided discovery and learning that can provide a heightened sense of agency. This is especially true for members of marginalized groups who participate in the cultures of computing and entrepreneurialism often promoted at hackathons. If hacking is to be not merely a site of activism that gets co-opted and absorbed back into corporate cultures of innovation, where hackathon participants are implanted with "el mismo chip," where they're asked to "add value" by translating injustice into corporate products and services, then the compacted space might serve to forge new solidarities, oppositional politics, or even calls to dismantle the

---

21. See Beltrán, 2017c.
22. Söderberg and Delfanti, 2015.
23. Irani, 2020.
24. Ames et al., 2018; D'Ignazio et al., 2016.

structures central to the oppression and dispossession of others.[25] As events continue to be organized by communities with a critical eye toward the potential for sociotechnical infrastructures to be deployed in the name of re-shuffling hierarchies of power and expertise, it's important to understand how hackers from other(ed) positionalities not only extend the genealogies of hacking, but also re-establish and re-orient them by imploding popular terms such as "hacking" and "entrepreneurship"—they learn to think against the ideology of "todos con el mismo chip."

One of the ways I point to the shifting tenor of hacking communities in Mexico is by using the term *hacker-entrepreneur* to identify my research participants. Though in subsequent chapters I analyze the ways a minority of them more tightly control the hacker identity, this hybrid term points to the way many often shift between these labels, or in some cases see no difference between the labels at all. A popular conception of the hacker might refer to someone who loves to program computers in the spirit of playfulness and exploration, and a popular conception of an entrepreneur might reference someone immersed in capitalistic or technocratic motives. A smaller subset of hacker projects, such as radical hacker collectives or *hacktivists*, were directly antagonistic to capital and sought to enact social justice politics, but hacker-entrepreneurs didn't conform to these strict demarcations. The term points to this fluidity between identities but also to the ways techno-entrepreneurial Silicon Valley–esque cultures have come to dominate hacking cultures in Mexico and Latin America more broadly. Some collectives might have originated as free software development and advocacy projects in the late 20th century,[26] but (whether they like it or not) they are now tightly coupled with the technologies and infrastructures that have emanated from techno-entrepreneurial cultures. Still, my research participants proudly wore the hacker badge across diverse spaces and collectives, usually to reference the fact that they were able to put in the code work, that they had the ability to immerse themselves in the code worlds and program computers.

One way research participants took it on themselves to hone in on what exactly a hacker identity symbolized was by developing hacker ethics. As programming extended in time and space from the confines of the

25. Irani, 2019. Maxigas (2012) makes a distinction between a later generation of "hacker-spaces" and the early "hacklabs." The latter were situated in anti-capitalist movements and barriers of access to communication infrastructures, which made them more overtly political in their call to spread access to the dispossessed and championing of folk creativity.

26. Kelty, 2008.

hackathon to the first hacker school in Mexico (and Latin America), for example, hacker-entrepreneurs gathered to improve their programming skills, work collaboratively on projects, and cultivate the hacker ethos by making explicit the ten principles of their "hacker ethics"[27]:

<1>   Give before you get
<2>   No pedir permiso (Don't ask for permission)
<3>   Hacer > Hablar (Doing > Talking)
<4>   No existen excusas (No Excuses)
<5>   Resolver problemas (Solve problems)
<6>   Sigue tu curiosidad (Follow your curiosity)
<6.2> Fracasar == Crecer (Failing == Growing)
<7>   Conoce tus herramientas y comunidades (Know your tools/communities)
<8>   Siempre aprender (Always be learning)
<9>   Involucrarse (Get involved)
<10> Divertirse en el proceso (Have fun)

The emphasis on self-sufficiency, curiosity, and fun espoused in this ethic is fundamental to defining hacker culture and identity, as well as to guiding hackers as they move through other domains of their personal and professional lives. The listing out of these tenets takes inspiration from the lore of hackerdom, especially other rules and manifestos that frame hacking as encompassing alternative practices and means of exchanging knowledge; modes of cultural and technical production that defy convention; counter-cultural politics; and most saliently, computing expertise.[28]

Hacking, especially when considered from outside the Global North, can become a site where young people work to either break *out of* or *into* global techno-cultures.[29] In contemporary Mexico and across the US/México techno-borderlands, young people learn to do both as cultures and imaginaries of "innovation" and "disruption" collide with practices of

27. Morato (2015a, 2015b) created both the English and Spanish versions of this list.

28. Levy ([1984] 2010), for example, provides a journalistic account of the founders of the Tech Model RailRoad Club at MIT, who took their technical curiosity to all domains of life, who saw themselves as hackers because of their shared interest in the computer as a revolutionary tool, and proposes six key tenets that guided their underlying set of morals, beliefs, and worldviews. See also Wark's (2004) hacker manifesto and Himanen's (2001) hacker ethic.

29. Nguyen (2016) works with Vietnamese hackers to show that people from marginalized locations might very well be looking to break into global techno-cultures from which they have been excluded. Nguyen argues that this is quite different than the breaking out of sociotechnical limitations that hacking in the Global North posits.

protest. With Silicon Valley and its corresponding techno-entrepreneurial culture always looming in the backdrop of many of these hacking spaces, emergent politicized forms of hacking can both accommodate and even succumb to market logics of competitiveness, agility, autonomy, and risk, while also providing openings for more critical, anticapitalist, and decolonial approaches. Considering practices of coding across the US/México techno-borderlands as an instance of "hacking in/from the South" further reveals the complex relationships between technology-driven capitalism, entrepreneurship, and hacking outside the Global North context.[30] The state builds more nodes and hackerspaces as it reinvents itself in relation to an ever-shifting modernity, but hackers create a collectivist response—within these spaces, they learn to hone their code work and use analogic thinking to "think with the code" and to slow things down, to inspect the iterative processes of exploitation and co-optation characteristic of global market forces as well as Mexican statecraft.

[2] Code Work AND the Ethno-Stack

To think with the code within hackerspaces means to think with but also beyond the "todos con el mismo chip" mentality espoused as part of the Mexico Conectado reforms. This government slogan, as well as the hacker-entrepreneurs' own moves with collectives such as the hacker school, made clear that coding as a symbol and as a metaphor to be applied to Mexican society was deeply contested between the hackers and the state. While government officials envision generic users plugged into computing devices—quite literally, as photographs of young people on computers is a favorite media trope—hackers not only develop their own specific ethics that guide their code work but, immersed in these worlds of computing, they learn how particular coding logics come to govern the ways people go about doing things in other domains—how the rhythms and movements that drive the code worlds deeply influence the government reforms many of them are against, or perhaps how they come to govern even their very own ideological commitments and everyday practices.

To trace how these logics and approaches from the code worlds influence life "outside" of them, we first must understand that what's "in the code" is

---

30. Amrute and Murillo (2020) propose "in/from the South as method" in hacking studies as a way of conceptualizing connections, divergences, and contradictions in how the Global North and the Global South hack and use computational technologies.

always influenced by more than the code. Matthew Fuller develops a vision and lexicon for a software studies which "aims to map a rich seam of conjunctions in which the speed and rationality, or slowness and irrationality, of computation meets with its ostensible outside (users, culture, aesthetics) but is not epistemically subordinated by it."[31] Indeed, a grounded and engaged approach to studying software should be attuned to the fact that what ends up in the code proper, as well as the very style and approach one brings to coding, is influenced by the dispositions carefully cultivated by conditions around us, and vice versa. In her longitudinal and multi-sited ethnography with FLOSS (Free/Libre and Open Source Software) hackers, Gabriella Coleman notes that because the technical craft of coding requires a constant awareness and rearrangement of form, hackers develop competence in transferring mental dispositions into other arenas of life.[32]

I use "code worlds" to point to the space and time coders inhabit when they become immersed in computer programming. I build on work by media anthropologists who have defined "media worlds" as "the network of production, circulation, and reception of expressive and audiovisual forms."[33] In the early 2000s, media anthropologists set out to understand "not only how media are embedded in people's quotidian lives but also how consumers and producers are themselves imbricated in discursive universes, political situations, economic circumstances, national settings, historical moments, and transnational flows, to name only a few relevant contexts."[34] By situating media as social practice within shifting political and cultural frames, ethnographic approaches promised to illuminate how media challenge (or enable) power structures, the imaginative alternatives to these underlying media infrastructures, as well as the role of media technologies in constructions of selves and collectives. While much of the early work on media worlds investigated how indigenous people used media technologies manufactured elsewhere as tools for self-representation, thinking with the code instead of with the media necessarily means thinking with the makers of the root infrastructures underlying technologies.

To explore the code worlds anthropologically thus means to consider how coding is inherently a social practice embedded in specific cultural and

---

31. Fuller, 2008: 5.

32. Coleman, 2013: 100.

33. Johnson and Jones, 2021: 7. For additional background on the disciplinary genealogy of the concepts "worlds" and "worlding" and how they have been applied to research with specific communities, see Ginsburg and Rapp, 2020.

34. Ginsburg et al., 2002: 2.

political contexts, influenced by power structures that mold the computing infrastructures themselves. To consider what it means to think or hack "in/from the South" means to treat the South not as a geographical south per se, but as a perspective that fights for the recognition of knowledges and forms of being seen as "other."[35] Guided by this approach and inspired by linguistic anthropology's contributions to media anthropology, which insists on not essentializing "digital communication" as separate from other channels of communication,[36] I explore how the social practices around the code worlds—what I call "code work"—take into account the production, circulation, and reception of narratives, artifacts, and subjectivities that arise when individuals and collectives navigate the code worlds. By staying committed to an anthropological focus on how difference and inequality are produced across domains of socio-cultural practices, and to further illuminate the back and forth movements between the code and social life, between the code worlds and what seem to be "other" worlds, I offer the *ethno-stack* as a framework to help us think with these traversals as well as to examine how difference is actively produced by and through code work.

The stack, in computing terms, refers to the interrelated and interdependent layers of hardware components and software protocols that make high-level computation and programs possible. To move from the bottom of the stack (e.g., machine code) to the top of the stack (e.g., programming languages and systems) means to traverse the corresponding circuits, microchips, and computer code that can be part of each layer of abstraction that makes up the system.

The fundamental idea is that one can navigate the stack by building up layers of abstraction from lower-level components. At the lowest level we might have MOSFETs (metal-oxide-semiconductor field-effect transistors), the most common transistors that ultimately produce the 0 and 1 bits; these 0s and 1s are fed into logic gates used as Boolean operators (e.g., AND, OR, NOT); these logic gates are then assembled with other gates, which become components used by other programs; and ultimately these programs are used by larger-scale systems, which are used by other systems, and so on. Across different layers of the stack, each configuration of elements becomes a component to be used by other components. The corresponding internal implementation of each element is abstracted away and largely irrelevant to the other components that use it.

---

35. Santos, 2014.
36. Johnson and Jones, 2021: 6.

Social scientists who research new computing technologies and their aspirational promises have proposed that in order for marginalized populations to infuse their own worldviews and future aspirations into a system, they must fully participate in and be adept at navigating all layers of the stack. The structure of the stack inherently hides conditions, keeping its range of possibilities from view. Becoming a full-stack code worker might help to uncover the stack and its liberatory possibilities.[37] Jason Edward Lewis argues that only by fully and comprehensively infiltrating and navigating all layers of the stack can Indigenous people, or other communities systematically excluded from the power of computing, increase their ability to "make the technology speak in the way that they [we] desire."[38]

Learning to navigate the stack across all its layers is one way to think with and beyond computing-as-usual, to think of alternative stacks or of alternatives to the stack. Even scholars who push the stack and its metaphors to the planetary scale to think about *everything*, from human and nonhuman users to state governance to climates,[39] leave conceptual space for other stacks to emerge. "We need not one but many Stack design theories," Benjamin Bratton tells us.[40] His invitation starts with what he calls "The Black Stack," the "generic term for The-Stack-to-Come that we cannot observe, map, name, or recognize."[41] The idea is that this more elusive and future-looking stack is coming and we can have a hand in modeling it. If we cannot escape the stack, at least it's designed to be re-designed and re-made.

Transdisciplinary scholars have taken this call to heart as they devise models for alternative stacks and stack thinking. A "diversity stack" might be made up of different layers (identity, chronotope, multimedia, and multilingual layers) meant to foster a virtuous circle in which research on diversity could help shape technological innovation, and technological designs could help understand and act on diversity.[42] A stack focused on creating a "new nomos for the post-capitalist common" engages with three levels of sociotechnical innovation, virtual money, social networks, and bio-hypermedia to invent

37. In the world of professional software development, a full-stack developer is a programmer who shows interest and mastery in all facets and layers of software development. A common way to describe a full-stack developer, for example, is as someone who can write code for both the back-end of a project (e.g., databases, architecture, servers) and the front-end of a project (e.g., graphical user interfaces, web applications, mobile clients).

38. Lewis, 2016: 242.

39. Bratton, 2016.

40. Bratton, 2016: 300.

41. Bratton, 2016: 368.

42. Liu, 2020.

new "social algorithms of the common."[43] Even Earth systems monitoring, understood computationally, may be brought into stack thinking.[44] The design of alternative stacks might be better able to accommodate racialized epistemologies, ontologies, and fields of action.[45] My thinking of a kind of stack "in/from the South" is informed by this work. As Amrute and Murillo formulate in their proposal for computing in/from the South, *in* the South centers digital infrastructures ethnographically to understand how they construct politics and ways of producing knowledge; *from* the South "opens up the material, immaterial, social, and political aspects of computing to alternative forms of life and future realities."[46]

Correspondingly, the *ethno* in ethno-stack points first to the definition of "ethno-" as a particular culture or people, this notion of difference signaling the *different* stacks that can emerge from stack theorizing.[47] In order to build the Black Stack or the Stack-to-come, coders might have to "first imagine it in ruins and work backward from this as both a conclusion and a starting point."[48] They might have to destroy the stack in order to re-build it. But once they are actively re-building this stack, once they infiltrate the deepest layers of this corrupted, monocultural stack, they might encounter the "ghosts" that first built it, along with their corresponding epistemologies and ideologies.[49] These ghosts in the machine might represent "phantasms," or a combination of images and ideas that become codified and reified in computing systems and encompass a "sense of self, metaphor, social categorization, narrative, and poetic thinking."[50] To take this stack by the horns they need to infiltrate its deepest layers to displace those ghosts or phantasms that have instantiated themselves at its core level, the root stack layers, which might hold the stories, worldviews, and epistemologies that govern all the other layers.

To begin to ground the stack thus means to "bend" its deep technical structure in order to make it work for corresponding communities.[51] This is

43. Terranova, 2014.

44. Helmreich, 2023.

45. Lewis, 2016: 241

46. Amrute and Murillo, 2020: 9.

47. I'm not promoting the essentializing of any particular ethnos (ethnic group) and understand any ethno-racial category not as an indicator of preexisting difference but as colonially conditioned and shifting, based on institutional and interactive structures (see Rosa, 2019).

48. Bratton, 2016: 357.

49. Lewis, 2016: 246–7.

50. Harrell, 2013: ix.

51. Lewis, 2016.

particularly challenging for marginalized groups who have been proportionately excluded from educational access to the stack and from stack theorizing. Bratton views the "user" subject within the Stack as "a position that can be occupied by anything (or pluralities, multitudes and composites),"[52] and this user might be something beyond the human, "*Users* (e.g. human, animal, machine) view The Stack and that initiate chains of interaction (columns) up and down its layers."[53] This might work for unmarked subjects accustomed to seeing themselves as generic users within a system, but for subjects marked along some dimensions of difference, their preoccupation might be with just making it to the "human" category, or with not being noticed as an exceptional user. The average "user" (white, male, heteronormative) is historically applauded for bending, hacking, or otherwise skillful technical maneuvering. For the racialized, gendered, or sexualized user, or for the hacker in/from the South, these moves are frequently criminalized and always surveilled.[54] To understand the role of computing in Othering and vice versa, we must think of computing not only as a field of expertise and as a set of converging technologies, but also as a means of organizing and differentially valuing knowledge as well as a method for surveilling and categorizing groups of people and their knowledge practices.[55]

The *ethno* in the ethno-stack thus also refers to the *ethnographic* approach that can lead us to think in this more expansive way about computing and its code worlds. Like the systems engineer, the avid ethnographer also deploys a type of "systems thinking" as they set out to their fieldsites or when they return from their sites to do the theorizing.[56] They observe or reflect about (or from within) a particular system to find its internal logic. Whether it's an economic system, a legal system, an educational system, or a particular community, ethnographers are determined to find out how these systems "work." Depending on their disciplinary training and what their specific purposes are, they develop theories about how these different (perhaps autonomous systems) interact to make up "society."

52. Bratton, 2016: 376.
53. Bratton, 2016: 375.
54. Beltrán, 2022.
55. Amrute and Murillo, 2020.
56. Ilana Gershon (2005) traces the work of Niklas Luhmann to show how "systems thinking" overlaps specifically with anthropological concerns about difference, the global/local, agency, and reflexivity. Thinking with systems theory allows ethnographers to develop concept work related to how participants experience and analyze their relationships to social orders, many times constituted as systems, but also about how people navigate systems with different and sometimes contradictory dynamics or internal logics.

The ethno- and ethnographic in the *ethno-stack* thus work together to ground the stack, to ask how it might be inhabited, contested, accommodated, resisted, multiplied, situated, or *bent*. Throughout the book's chapters, I offer four basic layers that can help us think with the ethno-stack: the personal, the interpersonal, the sociopolitical, and the sociotechnical. At the personal layer, the ethno-stack invites ethnographers, as well as coders themselves, to ask questions such as: how do practices of coding contribute to constructions of self? At the interpersonal level: what type of opportunities for solidarity-making across markers of difference (race, class, gender, sex, disability) are available at hackerspaces or sites that promote practices of computer programming? At the sociopolitical level: how do state representatives push nation-making narratives and ideologies of modernity when they sponsor such events, and how do participants respond to these political initiatives? And finally, at the sociotechnical level: how might researchers, together with research participants, source concepts and metaphors to make analogic traversals across these layers of analysis? Especially with this last layer, I argue that the code work might become "good to think with"— for code workers and for ethnographers—not only about how one moves from the back-end of a software project to the front-end, but also about how sociotechnical systems construct and are constructed along markers of racial, gendered, and embodied difference.

As computer programmers we can become adept at navigating the code worlds, but the ethno-stack prompts us to consider how the code work might provide us with the tools to re-envision, re-imagine, and re-arrange complex social relationships in and out of the code worlds. In the context of contemporary Mexico and hacking cultures that shuttle across the US/ México border, the ethno-stack becomes key to uncovering the tenuously constructed yet fiercely imposed borders that make up the political economy of tech and code work, or the *techno-borderlands*.

[3] Techno-Borderlands

In her genre-breaking writing, Gloria Anzaldúa develops the concept of the "B/borderlands." For Anzaldúa, the lower case "b" borderlands refers to the geographic region separated by the geopolitical Texas/Mexico border on which she grew up, and the upper case "B" Borderlands encompasses the psychic, sexual, and spiritual Borderlands of her own embodied subjectivities resulting from oppressions experienced due to her culture, color, health, gender, sexuality, economic status, and especially her complex relationship to language.[57] The B/borderlands, then, "in both its geographical and meta-phoric meanings—represent intensely painful yet also potentially transfor-mational spaces where opposites converge, conflict, and transmute."[58]

My notion of the techno-Borderlands pays homage to, and also plays with, Anzaldúa's concept to examine the political economy of tech and code work as it relates to US–México politics and the corresponding subjectivi-ties of difference found apart from yet deeply influenced by the US/México border. The influence of the borderlands on US/México hacking cultures can manifest itself quite explicitly, as, for example, when collectives on both sides of the border collaborate to organize the "Migrahack," a mobile hackathon aimed at resolving issues related to border securitization, US/México migra-tion, and surveillance of undocumented communities explored in Chapter 5. The more subtle Borderlands, with its soft mechanisms for constructing differences across intersecting dimensions of other differences, manifests more implicitly across various sites I explore in the book: gender and sexu-ality (all-women hackathon vs. performances of masculinity in mostly all-male hackathon), class (those who can travel to US vs. those who can only mobilize their efforts within Mexico), nation, race, and ethnicity (mobili-zation of Latinidad and Mexicanness in both US and Mexico). To explore US/México hacking cultures means to follow the varied experiences of the hacker-entrepreneurs who work in Mexico, those who travel to the US, and the "Latinxs" and "Mexicans" from the US who travel to Mexico—as well as their physical encounters across these borderlands. But more importantly, it also means to trace the subtle difference-making that is enacted, mobilized, resisted, and reconfigured across the techno-Borderlands.

Two key ideas from Borderlands theorizing that offer openings to study the politics of code work in US/México hacking cultures are the concepts of multiple allegiances and of in-betweenness. Thinking with the Border-lands means to move "within, between, and among these diverse, some-times conflicting, worlds."[59] Anzaldúa herself frequently moved between conflicting personal, political, and professional worlds, but what connected these worlds wasn't just overlapping or intersecting identities, as Anzaldúa's work is usually interpreted as suggesting.[60] A steadfast resistance to clear-cut labeling, coupled with this interest in developing novel alliances, "makes Anzaldúa's work vital for twenty-first century social actors, artists, thinkers

---

57. Keating, 2009: 1.
58. Keating, 2009: 10.
59. Keating, 2005: 2.
60. Keating, 2005: 2.

and scholars. Her words challenge conventional views that lead to stereotyp-
ing, over-generalizations, and arbitrary divisions among different groups;
her writings open up new spaces where innovative, sometimes shocking
connections can occur."[61] Thus, community-making across borders of difference is not based on sameness but instead on commonalities. In the case of
hacker-entrepreneurs across the techno-Borderlands who are already adept
at navigating the layers of computing stack abstractions, the stage is set to
hone this code work in order to find parallels between the code worlds and
other worlds they might not have imagined, be it the world of the humanities
or social sciences, the world of activism, or new worlds altogether.

The call of the Borderlands is also about finding power in the in-
betweenness, in the borders themselves. Becoming a threshold person, a
border-worker, means to locate oneself simultaneously inside and outside
of group formations, and to live "in between overlapping and layered spaces
of different cultures and social and geographic locations, of events and
realities—psychological, sociological, political, spiritual, historical, creative,
imagined."[62] Bringing the power of belonging to multiple worlds as well as
the transformative potential of existing *in-between,* scholars have proposed
research frameworks where both ethnographer and research participant
shuttle between differing, incomplete, and multifaceted viewpoints that
offer more complex understandings of ever-changing social realities by navigating spaces characterized by tension, struggle, conflict, and ambiguity.[63]
The "thickening of the borderlands" prompts us to look for the border in
the oppositional subjectivities it engenders as people bring "new energies,
new frequencies, new orientations" to confront these borders away from the
border.[64] The multi-border analytic thus means that in order to understand
constructions of communities and selves we need to think of borderlands as
"hemispheric, plural, and multi-sited."[65] Mobilizing these theoretical paradigms, women of color immersed in both cyborg politics[66] and intersectional
perspectives[67] have proposed weaving "between and among" oppositional
ideologies to arrive at a way of moving they refer to as "oppositional

61. Keating, 2005: 3.
62. Anzaldúa, 2000: 176.
63. Rosaldo, 1989.
64. Rosas, 2016: 355.
65. Guidotti-Hernández, 2017: 24.
66. Haraway, 1991.
67. Moraga and Anzaldúa (1981), Combahee River Collective (1979), Crenshaw (1989), and
Oyěwùmí (1997) were among the first scholars to consider how race, gender, class, and other
markers of difference overlap and intersect.

consciousness," a "differential mode of consciousness functions like the clutch of an automobile, the mechanism that permits the driver to select, engage, and disengage gears in a system for the transmission of power."[68] These emerging perspectives invite us to think from the borders themselves as we engage with the machines and the systems in order to decipher ever-shifting structures of power and inequality.

Code Work further prompts us to "think with the code" and across the ethno-stack to develop new terms and structures that, instead of cutting across difference, never lose sight of the political-economic and thus compel us to think about how inequity is structured and re-structured across domains, how recursive borders are produced and reproduced on a number of scales. The seduction of the stack is precisely that it allows us to highlight the visible from the invisible, to illuminate the known from the unknown; its separation of software from hardware, interface from infrastructure, allows us to decide what to show and what to hide as we play with the interfaces that lead us from one layer of abstraction to another.[69] The computer and its stack of abstractions provide a condensation of power that is "radically alien to most human experiences of the world. It is this alienness that allows software, particularly at moments when one is attempting to understand its working or to program it, that engenders the delicious moments of feedback between the styles of perception and ordering, logic and calculation, between the user and computer to be so seductive and compelling."[70]

How to bring it all together? How best to leverage the seduction and power of computing, and yet mobilize Borderlands perspectives to challenge the stack, the foundation of the computing infrastructure that guides the imaginaries and the political economy of the programming work? The argument I unfold throughout this book begins by paying attention to the code work already going on across the US/México techno-Borderlands, to take my ethnographic cue from the metaphors, logics, and ethics that these othered hackers deploy across domains. They use coding concepts and metaphors such as "batches," "exceptions," and "loose coupling" to describe their fraught relationships with employers and government institutions, as I analyze in Chapter 1. They recognize the complex intersections between their hacker ethic and transnational migrant ethics of "hard work," as I unpack in Chapter 2. They mobilize underlying coding principles of iteration and

68. Sandoval, 2000: 57.
69. Chun, 2013.
70. Fuller, 2008: 151–2.

efficiency to attempt to make them work in other domains of their lives, as I trace in Chapter 3. They rethink participation infrastructures and build solidarity by inviting programming newcomers to align themselves with specific layers of the stack, as I learn by participating in the all-women hackathon, in Chapter 4. They transform the practice of prototyping and iterative code design used across the stack to interrogate and propose versions of a transnational Latinidad, as I explore in Chapter 5. And they repurpose the tech startup logic of the "pivot" to represent their selves and projects across the unevenness of Silicon Valley politics, as I demonstrate in Chapter 6. Across these diverse spaces and experiences, I show that US/México hacker-entrepreneurs use their code work to develop heuristics for analyzing the organization of entities and relationships among them, whether they are elements in a coding environment, actors in a political-economic environment, or acquaintances in their intimate social environments. While I argue that this code work happens at the sociotechnical layer of the ethno-stack, I propose that connecting this work across its other layers—the personal, interpersonal, sociopolitical layers of the ethno-stack—will guide us toward thinking more holistically about the computing stack.

To cultivate *border-code-workers*, those who can connect the border work with the code work, is to provide code workers with the tools to think within and across layers of the ethno-stack, as well as within and across the communities that make up the techno-Borderlands. Returning to the call of the Borderlands, border thinking provides us a framework to "make connections among seemingly disparate events, persons, experiences, and realities" as well as to build on Anzaldúa's "holistic activist-inflected epistemology designed to effect change on multiple levels."[71] I've included the Ayotzinapa incident in this introduction because it points to interconnectedness of seemingly disparate communities as well as to the challenging era of political violence under which my research participants were honing their code work. Unfortunately, the incident was hardly remarkable.

Since the late 2000s, thousands of citizens in Mexico had been murdered or disappeared, and the Ayotzinapa incident simply mobilized the multitudes of citizens to protest the corruption, impunity, violence, and long relationships to drug trafficking that had come to characterize state practices in Mexico. In the state's eyes, the normalistas were treated as collateral damage—as coming from a space of social backwardness that was

---

71. Keating, 2005: 8. This foundation builds specifically from Anzaldúa's epistemological process of conocimiento.

preventing the rest of the country from moving forward on a developmental scale.[72] But coworking spaces, hackathons, entrepreneurial initiatives, and neoliberal "reforms" were seldom differentiated by politicians across their campaigns, as they simply shifted and constructed new categories of people that were keeping Mexico from advancing their modernist nation-building efforts. Becoming a border-code-worker, an approach spelled out more fully in the coda, means learning to mobilize coding logics and analogic reasoning while also thinking alongside the institutions and systems responsible for reinstating the unequal opportunities that many times result in violence and death, iteration after iteration.

Recalling the missed connections between the three collectives aboard buses that opened this introduction, I want to underscore that the potential for connection across efforts, perhaps the potential to hone this border hacking, was there all along. If the ephemeral nature of the hackathon forecloses any slower, longer-term solidarity-making and coalition-building that might lead to meaningful politics, El Chico Partículas might have been able to show Javo and his startup bus buddies how to build community with one's paisanos. If any diversity advocacy stemming from hacking cultures is ultimately too narrow when it centers technology as an orienting concept, both the publics on the startup bus and the combi de la ciencia could have learned from the normalistas, who have fostered class consciousness and fought for agrarian justice while surviving a century of iterations of repression masked as "national development," how to hone more radical politics. On the flip side, Javo might have helped la combi de la ciencia or the normalistas gain resources and support for their projects while not losing sight of their ultimate goals. The possibilities are impressive, but the fact is that the normalistas ended up murdered or "disappeared," while Javo and El Chico Partículas received prizes and acclaim. *Code Work* ethnographically investigates those moments across hackerspaces and hacker lives where such potentials could have crystallized, unpacks why they didn't, and proposes how they might in the future.

[4] Ethnographic Border Work

Three connected origin stories led me to investigate emerging forms of hacking and tech entrepreneurship between key sites in Mexico and the US. The first occurred as part of the "Latinxs in/and Tech Initiative," which I founded along with the help of undergraduate students at the UC Berkeley

72. Mora, 2017.

Center for Latino Policy Research. As part of the initiative, we set out to decenter mainstream "diversity in tech" discourse by critiquing dominant "bootstraps" narratives (for putting the onus of responsibility on underrepresented groups and individuals) and instead foregrounding the practices and structures that sustained the tech industry's underrepresentation of historically racialized groups.

As we met with and worked to connect different actors from academia, industry, and community-based activist circles, we were repeatedly appalled by obvious mis-connections between US-focused and Latin American-focused efforts. For example, at one conference in San Francisco dedicated to Latinx entrepreneurship, it was striking to note differences in discourse across two back-to-back panels: one featuring US Latinx speakers unpacked many of the complex diversity issues that we also had been pointing to, whereas the other, discussing opportunities for people from Latin America, disregarded these arguments and uncritically forwarded the meritocratic model of tech entrepreneurship. At another conference, Latin Americans were directly pitted against US Latinxs, with a panelist praising Latin Americans for "taking advantage" of the opportunities in Silicon Valley that US Latinxs were failing to capitalize on, blaming them for a lack of interest or even their weak entrepreneurial abilities. My frustration inspired me to dig deeper into these missed connections, and hackathon events proved to be ideal research sites to begin an ethnographic research project.

Code Work draws on participant-observation and interview-based research carried out between 2013 and 2020 both in Mexico and in the US, just before a newly formed leftist political party regained power from Mexico's "revolutionary" party, which had governed for nearly a century. In this decade, the tech startup scene surfaced in parallel with hype from economic analysts who projected that Mexico's economy was set to emerge as the "Aztec Tiger,"[73] and programmers from different socioeconomic backgrounds and across nationalized and classed boundaries gravitated toward code work. In my ethnographic research, I attended over twenty hackathons (usually multi-day) and spent extended time in hackerspaces, co-working spaces, and at tech industry events.[74] Drawing on my own undergraduate

---

73. Popular media outlets consistenly announced that Mexico was undergoing rapid economic growth that would lead to an increase in standard of living.

74. Organizers and participants in Mexico variously translated hackathon to hackatón or hackathón, or didn't translate it, using the English spelling. I stick to hackathon across the book for consistency and easier referencing.

training in computer science, I was able to participate actively at these events and in these spaces, sometimes even serving as a "technology mentor," as I brainstormed and prototyped alongside hacker-entrepreneurs, coding their projects or providing feedback on their tech startup ideas.

I conducted over fifty open-ended, formal and informal interviews with research participants primarily in two cities in Mexico: Mexico City (one of the centers of tech startup activity) and Xalapa (a small city in Veracruz about 300 kilometers east of Mexico City, where the startup and hacker community was unexpectedly vibrant). These two cities provided access to hackers from different geographic and demographic backgrounds. Mexico City is a mega-city where individuals more freely perform bicultural identities and interact with foreigners; Xalapa is a smaller university city of about 500,000 people surrounded by small municipalities where much of the economy revolves around commerce and services and the major employers are government and universities. Although the sites I investigate in *Code Work* are mostly situated in Mexico, my ethnography is transnational in that I traveled frequently between Mexico and various US sites, mostly in the San Francisco Bay Area, sometimes accompanied by my research participants, and sometimes running into them at various tech related events and spaces.[75]

Across these research sites I was able to develop relationships with a surprisingly heterogeneous cast of characters with unique backgrounds, motivations, and experiences. Featured characters are critical of, some even disenchanted with, the state-sponsored or neoliberal technology initiatives meant to boost a flailing Mexican economy, but many are also willing participants in these endeavors. Kike wants to live the excitement of the hackathon every day and decides to help found the first hacker school in Latin America. Leo, a veteran coder who religiously immerses himself in the vast libraries and functions of various coding languages, is concerned that this school is effectively "selling out" by commodifying the hacker identity, and distances himself from other dilettantes such as El Pato, who divides his time more equally between learning the ways of the code worlds and learning the philosophies and the language of the tech-startup entrepreneurial worlds. Mariana can code with the best of them in these mostly male-dominated hackerspaces, but is also an eager organizer and participant in the first all-women's hackathon in Mexico.

---

75. In order to highlight "transnational" phenomena, scholars have proposed traveling back and forth between their research sites (e.g., Joo, 2012) or trying to produce "ethnographic simultaneity" (Zilberg, 2011) between their sites located in spaces contained by national borders.

Some code workers have the privilege to travel to or reside in the US, or move across the US/México border frequently for professional or personal reasons. Armios follows the transnational Migrahack event across the border to work on hacking immigration issues that he's always felt close to, and his proclivity to think with the code persistently and across different domains (or systems) leads him to connect his "coder's paranoia" to a "migrant paranoia." Jessee, a Latino startup CEO, mobilizes an iterative way of thinking based on software development methodologies to consider the prototyping of future companies, coders, users, and especially of himself. Hiro thinks he can capitalize on his knack for hacking everything by attempting to hack the dating scene in San Francisco, with very questionable results. And the boundaries between the code worlds and other worlds are once again held up for inspection when Estefy tells her boyfriend, Rodo, that she doesn't want "MIT in bed with them," pointing to an unwelcome algorithmic-efficiency approach to their sex life.

This eclectic cast of code workers thus think with coding logics and rely on technical idioms to make sense of all sorts of non-technical aspects of their lives: their predicaments as migrants, their career choices, their love lives, and more. Because some of these characters make appearances in different chapters, they are listed in "Appendix 1: Cast of Code Workers" for easy reference. These concise profiles include their corresponding fieldsites, age and educational or professional backgrounds, relevant geographic or national origins, pertinent biographic details, and in some cases miscellaneous lifestyle preferences. All names are pseudonyms except for public or government figures, and unless otherwise noted.

My second origin story conveys how I was able to "enter the field" in these two Mexican cities. I had conducted preliminary fieldwork during the summers of 2013 and 2014, in Mexico City and in Xalapa, respectively, by serving as a technical instructor in a 6-week incubator program run by MIT meant to train recent university graduates with the technical and entrepreneurial skills to launch viable tech startup companies. As part of the "Aztec Tiger" initiatives, this type of project was easy to materialize in Mexico, receiving overwhelming support from government and universities, as well as interest from eager students. Participants in the bootcamp had backgrounds in software development, business/marketing, visual user interface design, and closely related disciplines. Through these contacts, I started to become active in the hackathon circles in both cities and developed relationships with participants and founders of the hacker school in Mexico City. Eventually I was invited to spend time with hacker-entrepreneurs outside of the hackerspaces and got to know them on a more personal level. These research experiences allowed me to begin to understand how hacker-entrepreneurs maneuvered their way through hacking and entrepreneurship worlds and mobilized their code work across other domains of their lives.

The final origin story is nested inside the second, for it explains how I became the technical instructor in the MIT bootcamp in the first place. As an undergraduate Course 6 (computer science and electrical engineering) student at MIT, I was seduced by the code worlds yet often felt an outsider to them. An ambivalent coder, I took a job post-graduation as a traveling business technology consultant working in both the US and Mexico, helping diverse organizations to design and implement custom software solutions. A defining experience came when a Mexican client, a large multinational corporation looking to update their complex legacy systems, tasked me with serving as a bicultural and bilingual broker between the US consultants and the Mexican code workers. It became clear to me that what was at stake was less their technical infrastructure's inadequacy and more a clash of sociotechnical "cultures." Fascinated by these differences and motivated to study "culture" more systematically, I went from ambivalent coder to ambivalent anthropologist, but always remained an unwavering code worker.

[5] Chapter Overview

My chapters think with the language of hacker-entrepreneurs to demonstrate how their practices of coding connect with their constructions of self and negotiations of diverse sociopolitical realities. Throughout, I return to the ethno-stack to show how specific hacking practices, logics, metaphors, ethics, and imaginaries are mobilized across personal, interpersonal, sociopolitical, and sociotechnical layers.

Chapter 1, "Thinking with the System in México," introduces the concept of *code work*, with an in-depth look at how hackathon participants immerse themselves in the code that underlies the technologies that promise developmentalist change. Hacker-entrepreneurs re-mediate pessimism and guarded optimism alongside each "new" version of modernity staged by corresponding political parties. The code work extends in time and space from the confines of the hackathon to the hacker school, where they carefully cultivate a specific hacker ethic in relation to state and market demands on their time, especially as the tech startup scene surfaces in parallel to hype from economic analysts who projected that Mexico was set to emerge as the "Aztec Tiger" economy.

Chapter 2, "Becoming Chingón at the Hackathon," mobilizes an intersectional analysis of how transnational labor, class, and masculinity come together at hackerspaces to consider how (mostly male) research participants enact hacker imaginaries and ethics as they strive to become "chingones" at the hackathon. I argue that the hacker ethics thus intersect with, align with, and sometimes challenge ethics formulated from the life experiences of young men who have to cope with the contradictions of their coding skills being valued in some spaces and undervalued in others.

Chapter 3, "Code Work across Domains," considers how the subjectivities underlying the code work is welcomed, or not, in domains of life outside of hackerspaces. Hackers tell me stories about how being immersed in the code worlds influences the way they solve problems across various domains of their everyday lives. Some give examples of how elements they associate with coding (speed, efficiency) and the entrepreneurial worlds (contracts, ephemerality) infiltrate their dating and sex lives. While some of these stories seem lighthearted, I end with the story of an encounter with healthcare inequities in the US to demonstrate how these cross-domain subjectivities are always connected to transnational constructions of race and class.

Chapter 4, "Abuelitas as Infrastructure," focuses on an event advertised in Mexico as the "first women's hackathon in Latin America." Participants in the women's hackathon work to align themselves with cultures of expertise as they negotiate normative roles of gender and femininity. Caught up in nationalist pushes for productivity, hackers weigh competing pressures in deciding what counts as "wasting time" at different moments during the event. A surprise visit by abuelitas (grandmothers) at the end of the hackathon makes a strong statement about unrecognized and invisibilized labor. I argue that the women's hackathon in Mexico serves as reminder to not devalue traditionally domesticized or feminized labor, as well as an expression of female solidarity, and that thinking carefully with the "bottom" layers of the ethno-stack can lead to inadvertently thinking with its higher-order, closely coupled layers.

Chapter 5, "Making Latinx Makers," looks at constructions of global Latinidad within the Migrahack, a series of hackathons that took place in both the US and Mexico. I connect scholarship on prototypes and participatory models with work that addresses the constructions and mobilizations of Latinidad to show that *making* prototypes becomes a way of *making* hypothetical versions of a transnational Latinidad, helping Migrahack participants to think through—as members of this community—issues related to US/México relations, border security, and migration.

Chapter 6, "Pivoting across the Techno-Borderlands," examines how research participants think with and against "the pivot," a tech startup term that calls for changes to a product that might better align it with the market. I show how Mexican and Latinx hacker-entrepreneurs pivot their identities, their language practices, their presence and presentation of self as they reconfigure the market logics of agility, competitiveness, and risk to creatively combine them with logics of hacking characterized by reinvention, playfulness, and resistance. I follow the trajectory of Javo closely across the techno-Borderlands to show how his app ends up returning to politics in surprising ways, arguing that this happens because he not only mobilizes migration as a type of hack, but also focuses his code work at the root layers of the ethno-stack.

Code Work ends with a coda, "Working Code AND Working Futures," where I expand on my notion of border-code-workers of the future, building on proposals from different media artists who have been inspired by border hacking; between "migration as a hack" and "hacking the border," I offer ways that the code work might be connected to the border work, and put forward questions that can guide our thinking with the generic ethno-stack, thereby grounding the stack logic of a variety of sociotechnical systems, present and future.

[1] Thinking with the System in México

Le dedico este retablito a Python el dios de los 0s y 1s
que después de tanto esfuerzo y desvelo mi programa
compiló justo a tiempo para la entrega.

Visionudo, Chicoloapan Edo. México

///ENG
I dedicate this retablo to Python the god of 0s and 1s
because after so much effort and sleeplessness my program
compiled just in time for the submission deadline.

Vision man, Chicoloapan State of Mexico

[0] The First Hack of the Day

Dozens of young tech enthusiasts wait in line to be allowed admission to the 2015 Hack CDMX event in Mexico City. Like other hackathons, this event proposes that participants show up, network, build a multidisciplinary team, and create a technological solution to a pressing societal problem. The winners receive cash prizes and a promise from the city government to provide institutional support for the project to be successfully implemented. The event is sponsored by over thirty government entities; if the heavy government involvement is somehow lost on any of the participants, they are promptly reminded when a caravan of black Chevy Suburbans pulls up to the building. Several square-shaped men wearing suits, dark sunglasses, and earpieces jump out of one of the vehicles and form a pocket around a slimmer man with a nicer suit as they approach the entrance. "Con esos lentes no pueden ver que hay una cola" (With those glasses, they can't see there's a line), one young man exclaims. "¿Quién es?" (Who is he?) I ask. "No sé y no me importa" (I don't know and I don't care), he responds.

Waiting in the line that curls around the street corner, a couple of young men spot an obscure door with a sign that reads, "Tocar en la siguiente puerta -> o la cortina de la vuelta" (Knock on the next door -> or the curtain around the corner). One of them quickly gets out a marker and makes two small modifications—adding a line to the "o" to make it an "a" and finishing the "r" to turn it into an "n"—to turn "cortina" (curtain) into "cantina" (bar). "There it is, the first hack of the day" (Ahí está, el primer hack del día), he announces to an approving crowd.

The attitudes and positions expressed in these brief interactions define much of the spirit and tone that will make up the weekend event. That is, these young people exhibit a sensibility for modifying, tweaking, and finding ways to exploit vulnerabilities in systems and structures, from the text on the sign to the practices of corrupt police officers. They embody and perform an ethos of "hacking" everything.

This hacking ethos will continue to be cultivated at the event, as it will be needed to participate fully at Hack CDMX. In a short window of time— forty-eight hours, the time displayed as a countdown on a giant screen overlooking the space—these enthusiastic programmers, entrepreneurs, designers, and community members will have to pitch their projects to over one thousand participants in attendance.

Waiting in line with me is Chavita, the top scorer in a software programming placement exam we administered as part of a summer-long tech startup boot camp at the Universidad Nacional Autónoma de México (UNAM) in 2013, where I served as technical instructor. In addition to hacking away on his computer engineering coursework at the university, he heads the university mobile development team, and during his free time Chavita performs the duties of "sensei" at Dev.F, a hacker school where young people gather to improve their programming skills, work collaboratively on projects, and promote the "hacker" ethos. At Hack CDMX, Chavita will continue to work in this spirit as he teams up with other members of Dev.F to develop Bikingos, an augmented-reality game that allows users to gamify their experiences using Ecobicis (Mexico City's urban public bicycle transport program).

Last year, Chavita's app, Audivio, won second prize in this competition. It used a crowdsourcing platform to help find missing persons in the city. Despite the city's promise to help fund and support the project, nothing materialized from Audivio other than a congratulatory letter signed by a city official and some winning pictures and press. The hackers know that In/fracción, EseTaxi, and ¡Aguas Güey! (other projects that will be developed at Hack CDMX) are likely doomed to the same fate. This isn't particular to this hackathon, or even to Mexico. Lilly Irani chronicles a similar experience at a Delhi hackathon: years go by without her demo spawning any projects, grants, or working systems, even though a team of talented professionals spent a grueling week developing software to create a sophisticated working demo. As Irani mentions, many hackathons have similar endings, where participants "just shake hands and say goodbye," and where much of what gets built "never gets built at all."[1]

Chavita, as well as other hackathon participants, are well aware of these dynamics of making and not-making. I asked Chavita why he showed up again this year to the hackathon, in the face of the same empty promises, but he responded with a reserved shrug. The examination of the underlying reasons why Chavita and other self-identified hackers continue to show up and help stage and perform the hackathon, in a setting where the promises of rewards and opportunities are largely spectacle, is one of the primary probes that guides this chapter.

1. Irani, 2015: 804.

This chapter has three main sections. First, I provide the political context in which hackathons and coworking spaces have been promoted by the Mexican government. To further explore the tension between the state-constructed hacker and the hacker who constructs their intervention, I delve into the social dynamics and software development practices within hackathon spaces. By focusing on the heterogeneous group of research participants that gravitate toward the code, I highlight how hacker-entrepreneurs navigate seemingly contradictory spaces and subject positions to understand their current sociopolitical configurations. The last sections dive into the code work itself; I argue that underlying design principles of the code such as "loose coupling" and fundamental software concepts such as "batches" and "exceptions" provide heuristics for analyzing the organization of entities and relationships between them. Between the code worlds and social-political worlds, the code work gives hacker-entrepreneurs the tool kit for social critique; it gives them the tools to think with the system, whether that system is the latest software infrastructure, socioeconomic program, or political reform.

As each iteration of hacker-entrepreneurs pushes forward their hopes and desires, they re-mediate pessimism and guarded optimism alongside each "new" version of modernity staged by corresponding political parties. Ultimately, Chavita and others keep coming back to the hackathons not because they're necessarily interested in creating a sellable app but because the events enable self-making. This making and re-making of selves is possible because the hackathon provides them an opportunity to experiment with and understand their current sociopolitical reality. Guided by their intimate knowledge of Mexican institutions, the hackers use coding logic to provide a foundation for honing their ability to manage themselves and their practices in order to scrutinize their relationships with the state, private companies, and their valued hacker communities. As members of this younger generation in Mexico negotiate their new subject positions and conditions, they carefully cultivate mindsets, attitudes, and strategies while they're "making," regardless of whether the apps and systems being made are ever actually made at all.

[1] Mexican Hackers as Model Entrepreneurial Subjects?

During the early to mid-2010s in Mexico, the tech startup scene surfaced in parallel to hype from economic analysts who projected that Mexico was set to emerge as the "Aztec Tiger" economy. Popular media outlets announced that Mexico was undergoing rapid economic growth that would lead to an increase in standard of living. My research overlapped with Enrique Peña Nieto's presidential tenure. During this time, his "reforms" were aimed to move Mexico "beyond low-wage factory jobs and toward an entrepreneurial economy."[2] Developmentalist narratives claimed that Mexico was "producing graduates in engineering and technology at rates that challenge its international rivals, including its No. 1 trade partner, the United States."[3] University enrollment in general had tripled in 30 years to almost 3 million students who wanted to join Mexico's "growing middle-class."[4] According to popular discourse, Mexico had become a top producer of "raw engineering talent," but it lagged far behind in basic measures of innovation, such as number of patents, scientific papers published, and research and development investments.[5]

Following these developmentalist narratives, Hack CDMX and other hackathons and coworking spaces fit into the larger Mexican political-economic landscape as potential generators of companies that might create jobs for them and their colleagues, and as the type of infrastructure that could help Mexico emerge as an entrepreneurial economy on the global innovation stage.[6] The tech-entrepreneurship movement in Mexico indexed the government's interest in rapid economic development defined by technical expertise and information technologies. Not surprisingly, state government offices could be found on the first floor of iLab, the original node that was part of the "Mexico Conectado" initiative and one of the coworking spaces where I conducted research in Xalapa. Politicians frequently went upstairs to hear the latest startup pitches and take pictures with the teams.

Here they visited 23-year-old Azukita, one of the many young Mexican entrepreneurs working feverishly at iLab to develop their ideas that might resolve pressing societal problems with the use of technology, but also with the hopes that their startup companies might attract the attention of venture capitalists or startup accelerators. Her team worked on Re-Active, a mobile

2. Booth, 2012.

3. Booth, 2012.

4. For an analysis of how middle-classness has become a powerful category for self-identification as well as the reference for ideal subject-citizenry, see Heiman et al., 2012.

5. Newspaper article titles such as "The Comeback Kid," "Mexico Makes It," and "The Rise of Mexico" give a sense of the cheers for Mexico's sudden emergence on the global stage (see, for example, *The New York Times* article by Friedman, 2013). *The Economist* (2012), for example, predicted that by 2018, "Made in China" would become "Hecho en México" (Made in Mexico).

6. Uribe (2021) further explores the construction of makerspaces and fab labs in México in the late 2010s under the government slogan "Todos somos creadores" (We are all makers).

platform that combined thermo-therapy and electro-stimulation to allow users to reduce chronic body pain using their smartphones. They presented Re-Active at the 2014 "Week of the Entrepreneur" event in Mexico City, a week-long national event aimed at bringing together Mexico's investors, politicians, tech community, and young entrepreneurs to boost Mexico's tech startup ecosystem. Ideally, their startup idea might scale to become Mexico's Facebook, Dropbox, or Amazon, the iLab rhetoric told them.

As recent university graduates with training in engineering, design, and business brainstormed to come up with the next business technology innovation, they did so under the gaze of model entrepreneurs such as Mark Zuckerberg, Steve Jobs, and Jeff Bezos, whose images and inspirational quotes could be found decorating iLab's walls. Together with witty tech wisdom such as "Just Google It" and writeable walls where anybody could start brainstorming startup ideas on a whim, the modern, playfully designed 4-story building opened its doors in January of 2014 with tech "innovation" and "disruption" in mind; iLab was designed as a space where young entrepreneurs, in teams of 2–6, could quickly generate, validate, and take their tech startup ideas to the market, where they might generate revenue (and perhaps also "make the world a better place") by "disrupting" the way a particular industry works.

But the young entrepreneurs by no means ignored the political backing and presence. Lotar, a self-identified hacker from Xalapa, comments on the upcoming hackathon being sponsored by the city government:

> "No pregunten cómo pero conseguí un borrador de la convocatoria para el "Hackatón Xalapa," por si quieren participar se pueden ir preparando. Hay que desarrollar soluciones tecnológicas que resuelven problemas de movilidad y servicios municipales . . . según esto ganes o no tienes que entregar tu proyecto (códigos y documentación) y pues el premio es salir en el diario de Xalapa, una beca iLab y una palmadita del presidente municipal . . . muy tentador no?"
>
> (Don't ask how but I got a draft of the call for the "Xalapa Hackathon," so if you want to participate you can start getting ready. We have to develop technological solutions to resolve mobility problems and municipal services . . . according to this [announcement], whether you win or not you have to submit your project (code and documentation) and well the prize is appearing in Xalapa's daily newspaper, a scholarship to iLab and little pat on the back from the municipal president . . . very tempting right?)

Lotar's sarcastic tone, in this case referencing the *palmadita* he would receive from the municipal president as a prize, points to the general skepticism and resentment toward government control of the hackerspaces and hacking events expressed by hacker-entrepreneurs. Across my research sites, they agreed that government entities were using hackathons (and hackers) as a way to further their own political agendas and as photoshoot opportunities for their poster politicians.

The use of technology by state entities to advance underlying political agendas is not new and has been theorized in other contexts. "Techno-politics" refers to a mode of politics that functions through invisibility.[7] Grounded in liberalism, governmental bodies seemingly leave citizens be to go about their everyday affairs without intervening. Instead, governments seek proxies in technological regimes—building sewers and other infrastructure, such as networks and phone lines, and conducting censuses—which are seen as technical and outside of political processes. Brian Larkin found a similar pattern in Nigeria: technology is used as part of political rule, and state-sponsored projects—roads, bridges, radio, any "new" technologies— are linked to events, to spectacular rituals that are meant to produce particular affective responses.[8] Not only are state officials always present in the mediated representations of these projects, before and after they are built (and even if they never are!), but the repetition of this pageantry in films and across different media is meant to produce, address, and train a modern subject on how to react to these awe-inspiring projects.

Likewise, the construction of the coworking spaces across Mexico worked to perform the potential of technological infrastructures to fulfill the promise of progress. If there were government offices on the bottom floor of iLab, and state representatives frequently visited to take pictures with the hackers diligently working above, it was because they wanted to be close to the modern, global subjects. These code workers were confirming Mexico's "coming of age," while governments across Latin America raced to move their nations away from categorization as "developing countries."[9] More importantly, the celebration of entrepreneurial hackers working within these spaces helped to promote a political agenda where young people were asked to appropriate neoliberal discourses about taking initiative, being self-satisfied, not waiting for government, and being "socially conscious."

7. Joyce, 2003; Mitchell, 2002.
8. Larkin, 2008.
9. Dávila, 2016.

The model entrepreneurial hacker emerged as a valuable subject in the Mexican political-economic landscape, where the majority of young people were disconnected from institutional support. That is, only a minority of young people in Mexico were connected to institutional circuits that allowed them to make decisions about their livelihoods regarding their health, work, education, and security.[10] A 2010 study showed about seven million young people aged fourteen to twenty-nine in Mexico were either looking for employment, not enrolled in school, or fell under the broad category of "not economically active."[11] This label is accompanied by a more colloquial term, *ninis*, short for *ni estudian, ni trabajan* (don't study, don't work), often used to refer negatively to young people who have become a burden to society or to the economy.

This labeling has its counterparts in other parts of the world. Vered Amit and Noel Dyck highlight the bureaucratic category of NEET (not in employment, education, or training) used to describe British youth.[12] The authors point to a specific category of youth all over the world who have been affected by economic restructuring. These young people are just as likely to be framed as agents of social change and progress as they are to index social breakdown and a "problem" in society. Young people in Mexico, particularly those from working-class backgrounds, express skepticism that university degrees will lead to gainful employment.[13] Many prefer to work within community- or university-sponsored spaces before participating formally in government and formal politics.[14] In interviews with youth in Mexico and Spain, researchers found that there was widespread identification with a "generation of disenchantment," as when research participants stated, "They fooled us, we did what they told us and in the end things aren't the way they told us they would be."[15]

Indeed, Lotar and other "disenchanted" youth proved important to the history of politics in Mexico. In fact, though, they proved to be less apathetic than one might imagine and instead followed the lead of the youth organizers who "engaged in the everyday work of creating social change, and through their collective political and social projects they place electoral politics neoliberal policies on trial for failing to offer meaningful solutions, hopeful

10. Reguillo, 2010; Valdez, 2009.
11. Instituto Mexicano de la Juventud, 2010.
12. Amit and Dyck, 2012.
13. Howell, 2017.
14. Ricuarte, 2018.
15. García Canclini and Cruces, 2012: viii.

futures, and channels for substantive participation for youth."[16] Many of these youth were thus central to the election of left-leaning Andrés Manuel López Obrador in 2018.[17] Running under the banner of the newly minted party Movimiento de Regeneración Nacional (MORENA), López Obrador won the most votes at the presidential level in Mexican history, an incredible thirty points more than his closest contender, and triumphed in thirty-one out of thirty-two of states and in 80 percent of the country's municipalities. López Obrador's political discourse, however, seldom had to give specific names or give detail of his plans for change. An anti-establishment stance against the *mafia del poder* (mafia with power) and critique of the "system" was enough to enamor young voters who held skeptical hopes for something new. Surprisingly, López Obrador's victory and promise of renewal came with familiar faces. That is, old adversaries had now become MORENA members, and López Obrador accepted everyone willing to join the party, with little gatekeeping.[18]

Larissa Adler Lomnitz, Claudio Lomnitz Adler, and Ilya Adler (1990) highlight how presidential campaigns in Mexico are carried out as political rituals in order to continue the same structure and organization under the guise of different campaigns.[19] Even before the parties' presidential candidates are chosen, there is tactical maneuvering and strategizing between *hombres del presidente* (men of the president) and *hombres del sistema* (men of the system).[20] The distinction between these two groups is one that reveals which subjects will emerge as the "politicians" (those who will maintain close relationships to the future president) and which will emerge as the "technicians" (those who will ensure the system continues).

Campaigning events are part of this political performance, for example, where organizers demonstrate how efficiently they can execute a flawless event; it's their opportunity to demonstrate to the presidential candidate and

---

16. Magaña, 2020: 10.

17. Villegas, 2018.

18. This wasn't entirely surprising, as López Obrador himself started his career as part of the Partido Revolucionario Institucional (PRI, or Institutional Revolutionary Party). From 1929 to 2000, the party held power while it constructed itself around the ideals of the Mexican Revolution and reinvented itself until it became representative of what many scholars call "institutionalized revolution." In 1929 it was formed as the Partido Nacional Revolucionario (PNR), in 1938 it was dissolved and renamed Partido de la Revolución Mexicana (PRM), and in 1949 was dissolved and renamed PRI.

19. Adler Lomnitz et al., 1990.

20. Terms used by Adler Lomnitz et al., 1990.

party how much they belong in the privileged "political class" made up of the "men of the president" and "men of the system." These stagings have another important purpose: they frame the "problems" that need to be resolved and stage the corresponding "dialogues" with special interest groups. We thus see the emergence of "women," "Indigenous people," "youth," and their projected prominence within these events indexes the extent to which the campaign will align these groups with the "problem" to solve.[21]

Thus, from Peña Nieto to López Obrador, "disenchanted youth" from all walks of life, especially those who were "not economically active," or *ninis*, were constructed as a problem to be solved by corresponding campaigns. "Technology for young people" is a pitch that translates across borders, including those that divide politicians and political campaigns. As "hacker-entrepreneurs," these young people were constructed as subjects who could themselves solve their problems, and the code would be a key element for them to do so. State-sponsored hackerspaces have endured from Peña Nieto to López Obrador, as "men of the system" and "men of the president" reorganize themselves and their projects. Within these spaces, young people were charged with demonstrating they could become the modern affluent knowledge workers of tomorrow.

But not all hacker-entrepreneurs in Mexico were necessarily driven by a naive interest in exhibiting the entrepreneurial spirit to perform middle-classness, to confirm their trust in their ability to build the nation and "change the world." I found a surprisingly heterogeneous cast of characters, motivations, and experiences at hackathons during a period of transition, or "disenchantment," in Mexico. Like the diverse set of constituents responsible for electing López Obrador, those who gravitated toward the code came from distinct social and class positions. To explore further why Lotar and Chavita, like many of the other coders at the Mexico City event, continued to hack away amid "politics as usual," which sometimes meant unreliable government sponsorships, shameless politicians, and empty promises, I participated as a team member and floating mentor across hacker events and spaces in order to stay close to their practices, hacking while I hovered above their hacking, within and beyond the hackathon.

21. Magaña (2020) further analyzes the politicization of youth identity in Mexico and the label of "youth" as a dynamic, relational, and heterogenous social identity that is historically constructed.

[2] Staging the Hackathon

If the Mexican state is invested materially and imaginatively in the hackathon, so are the participants who come to put in the code work to attempt to make their solutions and dreams come alive. The technical solutions proposed at this hackathon are closely guided by intimate understandings of the kind of apps that will win the event, that the "winning" might not lead to tangible solutions, and that there is real meaning in the process of "making" regardless of the outcome. Among the cast of characters at Hack CDMX is Leo, a veteran coder and recent UNAM computer science graduate who travels over two hours on public transportation in and out of Mexico City from a peripheral *municipio* to participate in these events. Most of his earnings he contributes to help pay for family expenses, and he saves up just enough to purchase airplane tickets to attend annual expo training events in the San Francisco Bay Area hosted by major tech companies. Wearing T-shirts given away by these companies is a badge of honor for the hacker, and they complement his wardrobe of more colorful T-shirts with even more colorful slogans: "Talk is Cheap," "Show me the Code," and Leo's favorite, "Programming is the closest thing we have to superpowers."

Another frequent hackathon participant who embodies the hacker-entrepreneur identity very differently than Leo is iLab's director, El Pato (The Duck). Other hackers gave him this nickname in response to his characteristic phrase, "Yo escopeta, tú pato" (Me shotgun, you duck), which is meant to index an overall disciplining of the rising hacker-entrepreneurs who join iLab. El Pato's bible is *The Lean Startup*, a popular book that circulates widely in the startup world and proposes a decentralized protocol for efficiently developing tech products that meet the needs of early customers, thereby reducing market risks and sidestepping large amounts of initial project funding. In an interview, El Pato tells me:

> "Lean significa esvelto, pero también significa que siempre estás en beta. Nada es seguro. Todo el tiempo estamos replanteando todo. Si ya estamos viendo que algo no está aplicándose correctamente lo calibramos. Si estamos viendo que hay algún proceso que deberíamos de estar adoptando porque está teniendo éxito en otra parte en ese momento lo conectamos con nosotros. No queremos hacer algo estático. Queremos que siga siendo muy dinámico."

> (Lean means slim, but it also means we are always in beta. Nothing is certain. All the time we are rethinking everything. If we see something

is not being applied correctly we calibrate it. If we see there is a process being applied successfully in another location, in that moment we adopt it and connect it to ours. We don't want to do something static. We want it to continue being very dynamic.)

When he makes appearances at events like the hackathon, he supervises iLabbers to make sure they are adhering to the disciplined entrepreneurship the model proposes.

Leo and El Pato thus lie on opposite poles of the hacker-entrepreneur spectrum. One was educated at UNAM, Mexico's top public university; the other was educated at the elite, private Tec de Monterrey. One sports the disheveled hacker appearance, complete with T-shirt and hoodie; the other prefers the sports-coat-and-jeans look characteristic of a Silicon Valley entrepreneur or investor. One religiously immerses himself in the code, with a vast knowledge of the libraries and functions that make up each coding language; the other splits his time between learning to code and learning the startup language of the entrepreneur. The playful nicknames and mocking serve to establish their distinct socioeconomic positions and differing roles within organizations such as the coworking space, but the fact is that Leo and El Pato end up occupying the same spaces, even if these appear to be at odds with one another in terms of politics or broadly defined cultural practices.

The ability with which different hacker-entrepreneurs also maneuvered between seemingly contradictory spaces was quite revealing during my research. Early in my fieldwork, I visited distinct hackerspaces to get a feel for the differing political alignments and ideologies across hacker communities. One collective, for example, more closely aligned with the images of hackerspaces a general public might have: undisclosed location, DIY workshops using open-source technologies, an alignment with radical leftist politics evidenced not only by discourse but also by Zapatista posters, and a collective community disdain for any potential member that pulled out an iPhone. This collective contrasted sharply with some of the corporate-sponsored hackathon spaces and events, complete with company-sponsored infrastructures to support software development (and sometimes gourmet catering), a reverence for Silicon Valley figures and discourses, and participants who compared specs of their latest smartphones. To my surprise, I wasn't the only diligent "ethnographer" navigating these seemingly contradictory communities; across these spaces, I frequently ran into my research participants, who plugged into and out of each collective with ease.

In the same way that "hacking" allowed event attendees to carry the ethos of the CDMX hackathon onto the streets (quite literally) in the opening vignette, the loose definitions of the term allowed Leo, El Pato, and others to come together across markers of difference within spaces that practiced distinct versions of hacking. In this sense, "hacking" functions as what Henrietta Moore calls concept-metaphors, whose meanings are necessarily underspecified to allow people (including anthropologists) to understand processes within specific economic and political contexts.[22] The purpose of these concepts is to maintain ambiguity in order to frame different levels or domains that facilitate comparison between contexts and open new spaces for future thinking. By traversing the different spaces that made up the hacker worlds, research participants worked on making the connections, comparisons, and contrasts that made up the communities that formed a specific cultural and political-economic landscape.

Membership in these hacker communities calls for the oscillation between craft and craftiness—the pursuit of quality and excellence as well as some degree of agility and guile.[23] Thus, as my research participants honed their technical craft, the underlying code work provided them the metaphors and tools to conduct this type of "ethnography," even if they never called it that. For hacker-entrepreneurs who spent a lot of time building systems, "hacking" provided a broad referent for navigating institutions, and the code work gave them the conceptual tool kit to carry out this social analysis and critique of other "systems." A closer look at this code work across my research spaces in Mexico reveals how the underlying logics of software design, specifically tied to the concepts of batches, exceptions, and loose coupling, provided heuristics for analyzing their precarious political and economic relationships.

[3] Code Work: Batches and Exceptions

At Hack CDMX, Leo, El Pato, and Chavita spend the weekend together thinking up solutions to Mexico's problems at the same time that they meet and work with other coders, designers, entrepreneurs, and curious onlookers from across the country and from across the world—participants that have shown up to take part in the spectacle of the hackathon. Hackers have gathered not only to create something new but to share, in person, their

22. Moore, 2004.
23. Coleman, 2017a.

latest creations; they show off their code to others who can appreciate it. Frequent exchanges of "Eres un chingón" (You're a badass) circulate as participants evaluate and confirm the conciseness, efficiency, and elegance of each other's code snippets.

Indeed, the principles of reuse, simplicity, consistency, efficiency, and the ability to shuttle between different levels of abstraction are core tenets of computer science and the metrics used to identify a talented computer programmer. Hackers at Hack CDMX use the time and space to share code from other projects they have been working on, sometimes from their professional jobs, where there are few programmers, and where results-oriented managers fail to recognize the complexity and beauty of their creations. In her ethnography with hackers, Gabriella Coleman finds their values of cleverness, ingenuity, and wit transfer to the process of making technology and writing smart pieces of code.[24] That is, hackers "revel in directing their faculty for critical thought toward creating better technology or more sublime, beautiful code."[25] If one can dissect, manipulate, reassemble, and solve the problem within the given constraints and tools at hand, one can create beautiful, "original" code.

While some newcomers learn the ways of the code worlds at the hackathon, others have carved out more permanent spaces to carefully cultivate their code work, along with the corresponding attitudes and values. "We loved to go to hackathons so we made one that would extend more in time. We wanted to live the hackathon every day," Kike tells me. He is one of the founders of Dev.F, the first "hacker school" in Latin America, created in 2014. As students at the hacker school progress from white-belt (most basic) to black-belt (most advanced) classes, their "senseis" (Chavita among them) provide feedback and mentorship into the hacker ways. They proudly wear T-shirts designed in-house, with their logo on the front, and their vision for how hackers fit into the overall political-economic landscape succinctly stated on the back: "México necesita más hackers" (Mexico needs more hackers). One of first entries on the school's popular blog lays out the ten principles of the "hacker ethic" one must follow to become a Dev.F hacker.[26] Rules listed in the hacker ethics, such as "give before you get," "conoce tus herramientas y comunidades" (know your tools/communities), and "no pedir permiso" (don't ask for permission) reference the values of

---

24. Coleman, 2013.
25. Coleman, 2013: 118.
26. See Appendix 0: Glossary for listing of principles.

self-reliance, mistrust in authority, and intense dedication that define previous hacker cultures that have inspired them,[27] but also the love for self-reinvention, the *"There is a better way, and I can do it myself"* belief that has driven a broad spectrum of attitudes and practices of techno-entrepreneurial cultures.[28]

This hacker ethic is thus carefully cultivated at the hacker school. It guides the mindset and approach to various life situations that hackers are supposed to value as they hone their coding prowess. It helps orient their constructions of self-identity and provides simple rules to revisit when they see themselves in a complicated, real world conundrum. These straightforward tenets that make explicit how one might embody the hacker culture are welcome by most in the hacker school.

A minority, however, feel that the "hacker culture" has become a product being sold to hackers-in-the-making by the expert coders who run the school. They're quick to call out that some of the language used in the hacker ethic aligns with neoliberal language that calls for young people to take matters into their own hands. For example, "<3>hacer > hablar (doing > talking)" and "<4>no existen excusas (no excuses)" call for productive, noncomplaining citizens.

Leo, who holds a strong conviction about what a hacker should or should not be, has related problems with the hacker school. "Their website even looks like the template all the startups use. What's the difference?" he asked. Leo does not agree with the "real" hacker ethos being compromised. He's expressing one of the fundamental tensions I heard hacker-entrepreneurs debate: the anticorporate hacker ethos versus the embeddedness and dependence of hackers on tech companies. The hacker school sells the hacker package to not only hackers-in-the-making but also the tech companies who want to integrate talented software developers and feel they are part of something young and new.

Kike tells me more about how the school helps burgeoning hackers connect with potential employers. Since the school is nomadic, he tells me, each batch of students takes part in the twelve-week program in a different part of Mexico City, many times holding their trainings and related events within coworking spaces. However, they also hold some trainings and events within tech companies and government offices. When they work within one of these facilities, they are not involved in the operations of the organizational

27. Levy, (1984) 2010.
28. Graham, 2022: 395.

entity, but they do promote hacker students for advertised jobs within the company or government office. The idea is that a few participants might transition into professional roles within the organization that do not compromise the hacker ethic the Dev.F program has so carefully cultivated. "If there's no match, the hacker comes back to next batch and we try again. It's like catching an exception," Kike tells me.

The metaphors that participants at Dev.F used to describe their practices were fundamental to their code work. A batch, for example, is simply a group of items. The term can also be used to refer to a set of instructions or processes that run before or after a user interacts with the computer, as in "batch processing" or a "batch job." In the case of Dev.F, each batch of students developed their coding skills and hacker persona before they were presented to the potential "user"—in this case, a company that would potentially find the coding skills of the hacker to be of value to their business. If there was no "match," as Kike described, or if during the time that Dev.F worked within the organization's space they deemed that their coding skills would not be truly valued, or that the entity might compromise their hacker ethics, then they would try again—in computing terms, they would "catch the exception." An exception occurs when there is an anomaly or unusual occurrence during the normal flow of execution of a computer program. A defensive programmer plans for these exceptions, thinking about anything that can go wrong during the execution of a program, and prepares special cases within the code to handle these exceptions. As each of their batches iterates through the hacker program and the school organizes events within tech companies or state-sponsored spaces, they interview workers to get a feel for the workplace dynamics (e.g., retention rates, coder satisfaction, nepotism) and the political moves within. Instead of simply responding to job postings and allowing companies to exploit their coding skills, Dev.F envisions that their careful analyses of the inner workings of the company will allow them to create an overall better system.

The time spent within the hacker school thus serves to resolve one of the fundamental predicaments that many of the young university students and graduates must work to resolve in relation to their studies and their labor. On the one hand, they could be unemployed or participate in intermittent or poorly remunerated employment; on the other hand, they could continue to wait until the ideal job comes along. Julie Archambault shows how graduates in Mozambique follow the latter route and wait it out until an acceptable job comes along. These youths are criticized by older citizens for their sense of entitlement, for preferring to go hungry instead of taking menial jobs; young people come to expect social mobility after years of studies.[29] Unemployed graduates in India similarly linger in a state of educated stasis, acknowledging that further studies might make them "overeducated," which might further increase their unemployability.[30]

While young people in Mexico also express ambivalence toward education (and overeducation), the hackerspaces give them the space to collectively socialize as they rethink their situation and plan their course of action. They also provide them a sense of safety in numbers—they can name themselves and their desires as being part of a collective, rather than as an entitled individual or simply someone who exists out of bounds of what is acceptable behavior. If "the state" inserts itself on the bottom floor of iLab to monitor the work of the hacker-entrepreneurs and use them as models for photo opportunities that promote the "promise of technology," as my research participants claimed, then the hacker school uses the coding metaphors to reverse the relationship. That is, the hackers attach themselves to entities to get a feel for the institution, to "interview" them while the hackers analyze if it's an institution they want to be committed to. Their work within these spaces might effectively help them dodge the problematic categorization of *ninis*; they inevitably perform the role of knowledge workers enthusiastically connected to their computers for various publics. But the real work, the code work, slips under the radar.

This code work effectively connects the code worlds to other social and cultural worlds my research participants inhabit. But it's not that "the code" is bounded from these other domains. Nick Seaver shows how teams of software developers organize themselves according to principles that govern the code they write; coders build small, semiautonomous, readily configurable groups.[31] In the same way Seaver pushes us to think about algorithms as complex sociotechnical objects instead of autonomous technical objects that can come to "rule" human culture, we can move from thinking about code as a purely technical object and thinking about the code worlds as sociotechnical fabric. My focus on the code work highlights precisely how this fabric is woven (or coded). The way analogies are drawn together between different domains is what anthropologists have referred to as a "culture."[32] If scientists and engineers encode their (many times Eurocentric and heteronormative)

29. Archambault, 2017: 79.
30. Jeffrey, 2010: 472.
31. Seaver, 2018.
32. Strathern, 1992: 33.

worldviews into computer programs,[33] here they use the very logics with which these programs are constructed as transformative openings that have the potential to make different worlds possible. For hacker-entrepreneurs in Mexico, fraught relationships with political and economic entities might be the reality of their existence, but batches and exceptions provide the metaphors and (hopefully) tools to think of these entities as elements in the current configuration, and to realign their own practices accordingly. The code work allows them to think and act alongside "the system."

[4] Code Work: Loose Coupling

With one hour left in the hackathon, the Bikingos team takes a needed break for some noncoding *cotorreo* (fooling around, just hanging). I take advantage to conduct informal interviews with the team. Leo tells me more about why he is so tired. Leo works for a tech consulting firm in Mexico City and usually spends ten to twelve hours a day programming, and he often has to work weekends with no extra pay since he gets paid by the project. His best friend Memo, who comes from a similar socioeconomic background, recently landed a job doing back-end coding for a major bank in Mexico. Memo now has a guaranteed salary perhaps worthy of his UNAM degree; the 17,000 monthly pesos (~US $850 in 2014) solidify his position as a lower-middle-class citizen who no longer has to worry about making ends meet. A monthly salary of 16,000–25,000 (~US $800–$1,250) was the norm for a recent software engineering graduate from a reputable program; for someone with 2–5 years experience the monthly salary jumped 30,000–50,000 pesos (~US $1,500–$2,500); and for a veteran programmer with 10 or more years of experience the monthly salary could reach close to 100,000 pesos (US $5,000). These potential earnings and job security make Memo happy and his family proud of him.

Leo, on the other hand, isn't impressed with Memo's salary, and much less with him "selling out." Although Leo lacks job security and makes less money working on projects for the consulting firm, he claims he'd rather work on the more challenging projects the firm hands him and that, even if the work was similar, they couldn't pay him enough to wear a suit and join the *godinez* (office workers) at the bank. He is aware of the precarity of his situation, but instead of framing it negatively, he refers to his work arrangement as "loose coupling."

33. Helmreich, 1998.

Loose coupling is a computing term that refers to a robust way to write code where data structures (or other components) can use other components in an interconnected system without needing to know the full details of their implementation. In this way, each component becomes more autonomous and can be used for different purposes by different components; elements become "coupled" and depend on each other with very little (or no) direct knowledge of each other. Leo goes on to recommend manuals and tutorials that further explain this software design so that I can appreciate its value. The term "loose coupling" Leo uses to refer to his flexible work arrangement references his autonomy at the same time that it references his replaceability. Like many of the young people in attendance, Leo contracts out his programming skills to diverse companies and startups. In his case, the consulting firm helps make these connections, especially with US-based companies looking for programmers who work for less pay than software programmers in the United States.

Leo changes gears in our conversation and elaborates on the hackathon dynamics: "In the past, the politicians would arrive to distribute blenders and take pictures when the basket[ball] court was completed, if it was ever completed. Now, they arrive to distribute hackathon stickers and take pictures with the winning teams" (Antes, los políticos llegaban a repartir licuadoras y a tomarse la foto cuando se terminaba una cancha de basquet, si es que se terminaba. Ahora llegan a repartir stickers del hackathon y a tomarse la foto con los equipos ganadores). The stickers Leo refers to are primarily used as marketing material: they show the logos of tech companies, operating systems, development tools, and hackathon events, and participants like to decorate their laptops with them, creating colorful, creative displays. Even though Leo criticized the practice of sticker distribution, associating it with the "old" method of gifting household electronic appliances such as blenders in the name of voter recruitment by politicians, he still proudly displays his stickers on his laptop. Moreover, the varied events, companies, and technological platforms show the contradictory and fleeting allegiances that currently make up his hackerworld. Like the "loose coupling" approach he takes to code, his sticker arrangement points to his flexible (and legible) networking capabilities.

Leo talks about the politicians (as representatives of the state) in the same way as companies, both as elements that can be represented visually and reconfigured on his computer. Instead of thinking of these sticker arrangements as pointing to the contradictory collectives to which the hacker-entrepreneurs belong, we can think of these shifting compositions as a canvas where (fleeting) relationships are made explicit and negotiated. "Loose coupling" gives him a way to think through these relationships. In the same way, the hackathon serves as a space where individuals negotiate not only their belonging to communities but their position within broader political and economic processes. Similar to the way Leo voiced his hesitancy with the hacker school "selling out" by adopting startup cultural forms, my research participants used the hackathon as a space to debate common points of contention: the desire of being associated with Silicon Valley infrastructures versus the reality of being exploited by them; the focus of autonomy and DIY central to the hacker ethos versus the promise of state care; the aspirations of hackers to benefit from their involvement with tech companies versus the practice of being exploited by them.

These shifting relationships and negotiations also point to the neoliberal knowledge economy and underlying processes of transnational capitalism that ask young people to work by the project, and on their own time. Rising hacker-entrepreneurs must learn to respond quickly and with agility to volatile market trajectories and frequently cross career, role, and political boundaries to perform their flexible or "latitudinal citizenship."[34] Market volatility becomes a way of life, where flexibility, instability, liquidity, and risk-taking are interpreted as desirable and challenges that the modern subject can manage by employing calculative decision-making.[35]

But unlike other entrepreneurs who embody the rhythms of neoliberal life (in Mexico or beyond), Leo and other hacker-entrepreneurs stay close to the code. Concepts such as "loose coupling" serve to make sense of the shifting relationships, even when these relationships are with the institutions that promise to help them navigate these relationships (e.g., the hacker school). Their lived reality calls for young people to be flexible, but not too flexible, autonomous, but not too autonomous. As Ilana Gershon reminds us in her study with Silicon Valley job seekers and job hoppers, enacting a neoliberal self is tricky. Workers come to see themselves as projects who must steer through various possible obstacles and alliances, moment after moment, with each instance creating a possible contradictory dilemma.[36] Neoliberal workers develop different strategies across different national and cultural contexts for "making do" in precarious labor markets. Young people in South Korea, for example, learn to "think with play" when they immerse

34. Ong, 1999.
35. Ho, 2009; Miyazaki, 2003; Zaloom, 2004.
36. Gershon, 2018: 175.

themselves in the digital gaming worlds and develop dispositions for think-ing and acting in quickly shifting, unsettled circumstances; strategies for "productive slowness" allow gamers to take ownership over their time and activities in irregular labor arrangements.[37] In Mexico, my research partici-pants might not necessarily "escape" neoliberal work conditions, but the code work allows them to slow things down, to take momentary snapshots of the way their economic and political connections are drawn together. The code work offers heuristics that might lead to transformative open-ings in volatile conditions as well as metaphors that provide the promise of autonomy as young people assume their positions as "neoliberal workers."

For forty-eight hours, then, Bikingos team members put in the code work as they design a beautiful graphical interface and user experience for their application and test their app while riding bikes around the city. After several iterations of prototyping, testing, and debugging, they commit their final code snippets to the team's repository, click "deploy," and celebrate the successful launch of their working application. They deliver a phenomenal pitch to the hundreds who show up for the final demo session and celebration. Bikingos wins first prize in the "solutions for the city" category, part of larger govern-ment efforts to revitalize the historic city center with "green" and "smart" technologies.[38] The team poses proudly for their group photograph. Chavita gets the same certificate he did last year. In the individual photo sessions, a different politician than last year takes a picture with him.

This public performance of the rewards and the potential of the hack-athon contrasts with the private discussion the team had as they talked about the actual utility of the app. That is, they were aware that the rating system and route-sharing infrastructure that was part of their app was not very likely to be used in daily Mexico City life; because of privacy issues, users would be reluctant to share any personal information, despite the promise of secure connections and encrypted data. Thus, hackers know that their code "works," in that their platform delivers the expected outcomes, but their final "product" might not necessarily be used by actual citizens. Whether the judges at the event realize this or not is many times beside the point. Regardless, what brings people who fall on different sides of political and institutional boundaries is the code; they're at the hackathon for the code work, not necessarily the "results."

37. Rea, 2018: 509.
38. See Leal Martínez, 2016; Crossa, 2009.

Precisely, a fundamental feature of the "making" these code workers are involved in is the understanding that noncompletion is a legitimate outcome of their efforts. Across maker communities, the language of "prototyping" is used to reference the experimental, open-ended, and often aspirational desires for communal self-organization embedded in skill-building.[39] That the objects being built are less important than the people and mindsets being built directly contests the language of emancipation and "access" that often accompanies the introduction of computers, especially in "underserved" populations or "developing" countries.[40] Whether the discourse and corresponding infrastructures are introduced by foreign companies or by the state, in this case, my research participants retreat to the lower layers of the computing stack—to the underlying computational principles and logics that structure the code—to help them make sense of their current situation.

Techno-politics are meant to distract citizens, to use technological proxies that appear to be outside of "politics" to continue business as usual. In Mexico, the hackathon serves as another ritual to perform modernity and the promise of technology. Software's fundamental appeal is that it has the power to illuminate the unknown; its separation of software from hardware, interface from infrastructure, provides a powerful metaphor for how the system works.[41] Young people adopting, constructing, and navigating these technical infrastructures feel less threatened or "trapped" by them and more inspired to "recapture them and turn them to new ends in the service of new worlds."[42] Between the code worlds and social-political worlds, the code work allows hacker-entrepreneurs to construct new forms of mobility by traversing the bottom layers of these "new" technologies, where they can observe how different elements are related, how things "really work."

[5] Still Waiting (in line for the hackathon)

When I initially asked Chavita why he continued to attend hackathons in the face of empty promises and uncertain outcomes, he responded with a reserved shrug. By following his and other hackers, moves in the world of hackathons, coworking spaces, and hacker schools, I was able to see how Chavita, Leo, El Pato, and other hackers actively participate across communities as they appropriate and embody the hacker spirit. That is, in some ways

39. Corsín Jiménez, 2014.
40. Ames, 2019; Crooks, 2018.
41. Chun, 2013.
42. Seaver, 2019: 433.

they belong to the undifferentiated "global" hacker community other scholars have conducted research with. They value cleverness and creativity and
place a high premium on knowledge, self-cultivation, and self-expression as
core tenets of achieving "productive freedom" and corresponding "software
freedom."[43] They improve their technical craft by following principles of
reuse, simplicity, consistency, efficiency, manipulation, and agility.

As these hacker-entrepreneurs honed their coding skills during my
research, multitudes of citizens across Mexico collectively protested the
impunity, corruption, and violence that had come to characterize state
practices, where *narcofosas* (drug-trade graves) with hundreds of unclaimed
bodies frequently appeared in clandestine locations and where dozens of
protesting students went "missing" at the hands of state officials. The case of
Ayotzinapa discussed in the book's introduction was a particularly dramatic
and mediatized event where 43 students were forcibly disappeared. Mariana
Mora argues that because these students were constructed as coming from
a space of social backwardness, preventing the rest of the country from
moving forward on a developmental scale, they were simply a casualty of
modernist nation-building efforts.[44]

Although many of my research participants came from similar socioeconomic backgrounds as the disappeared students, their activities within
the hackerspaces helped them dodge a similar fate. At one level, their
demonstrated activity within these spaces served to avoid their labeling
as *ninis*, in that their activity might be qualified as "studying." Moreover,
the genre of their studying fell under the appropriate category of respectability. That is, they weren't learning how to farm and becoming versed
in transformative pedagogies, as the Ayotzinapa students were doing; the
hacker-entrepreneurs were performing their roles as affluent knowledge-
workers-in-training within the modern, sanitized coworking spaces. If they
happened to come from marginalized backgrounds, they were demonstrating they were capable of breaking intergenerational cultural and economic
"backwardness."

As I've argued here, though, my research participants appropriated the
discourses of flexibility and self-management at one level, while they allowed
their code work to guide them at another. When faced with the developmentalist "reforms" of neoliberal economic change, middle-class youth may
respond by exploiting their advantages vis-à-vis the poor, or by building

---

43. Coleman, 2013.
44. Mora, 2017.

solidarities to protest the status quo. Among Mexican hacker-entrepreneurs, I found a heterogeneous cast of characters, motivations, and experiences; people who were not just driven by misguided interest in exhibiting the entrepreneurial spirit in order to perform middle-classness. I've suggested that youth from diverse socioeconomic backgrounds gravitated toward coding during a time of transition in Mexico, when a newly formed leftist political party gained power after a century of rule by a party characterized by unmet promises, a party that continuously shifted its "men of the president" and "men of the system" to give an appearance of change.

In this context, "hacking" emerged as a way for young people to make sense of their futures in a precarious state and economy, and as a way to let the "code work" intervene in narratives that have only delivered false hopes. Coworking spaces, hackathons, entrepreneurial initiatives, and neoliberal "reforms" are seldom differentiated by politicians. Hacker-entrepreneurs become part of the reimagining of Mexico as an orchestrated national project, entrenched in narratives that promote the promise of technology. By immersing themselves in the code worlds that underlie these technologies, they learn to design systems that promote separation of concerns and self-determination by actors. "Loose coupling," "batches," and "exceptions" provide heuristics for analyzing the organization of entities and relationships between them, whether it's an element in a coding environment or an actor in a political-economic environment. The code work offers a younger generation in Mexico the conceptual tool kit to think about political economy on the ground in the context of changing state power, shifting meanings of entrepreneurialism and capitalism, and a challenging era of political violence.

As these young people turn the spotlight less on what they say and more on what they code and the context in which they do so, they hack away, and in the background we have business as usual, politics as usual, reforms as usual. Politicians create and re-create "the state" in response to narratives that paint Mexico as "hyperconscious of its backward condition for at least 150 years,"[45] or as a place where "traditions have not yet disappeared and modernity had not completely arrived."[46] The hackathon becomes a site where "new" versions of modernity are staged, where the state and hackers find complex ways to coproduce themselves, and where coding logic becomes foundational for the reconfiguring of these relationships. Here,

45. Lomnitz, 2001, xvii.
46. García Canclini, (1990) 2009: 13.

the self-identified hackers find meaning in a community of action and performance that supports them as they negotiate their new subject positions and conditions within these overarching processes that construct them as always "in-the-making," as always "becoming," as always waiting. If they're going to be waiting, they might as well be waiting in line at the hackathon.

[2] Becoming Chingón at the Hackathon

Bendito seas Ángel Guardián del Hacking. Arreglé mi taller bien chingón y el jefe finalmente pudo ver que escribir código también es trabajo duro.

Cofi, 2015

///ENG
Blessed art thou, Guardian Angel of Hacking. I fixed up my workshop real badass and my pops was finally able to see that writing code is also hard work.

Cofi, 2015

[0] From *Carne*worlds to *Code*Worlds

Excitement brews as hackerspaces are constructed across Mexico and more young people create communities that align themselves with the hacker ethic. While young people flock to hackathons across the country, those who are able continue to visit Mexico City's hacker scene. Because of its size and status as a cosmopolitan city, it becomes the first point of arrival for global initiatives that begin in the US or other places in the world with a reputation of being hubs for innovation. Such an event is Angel Hack, which promises the winners a trip to Silicon Valley to pitch their startups to US investors. The crème de la crème of hackers show up to the event, not only to measure their coding abilities with other coders but also to potentially win the paid trip to the San Francisco Bay Area.

At the event in 2015, the ParkInMySpace team members spend the weekend putting in the code work to design an app Mexico City residents can use to rent out their parking spaces. The app's name isn't the only thing you'll see or hear in English at the event. English-language music of diverse genres emanates from laptop speakers from the different workspaces teams have designated for themselves at the hip new co-working space in a posh neighborhood (Condesa) in Mexico City. One will frequently hear participants integrate English-language words into their banter, as if to signal to the event organizers that they are ready for the potential trip to the US.

Leo and the other two team members are joined by another six young men from other teams, who enthusiastically gather around Chavita as he prepares to demonstrate to this select group how his code for ParkInMySpace works. Of about 100 participants at this Angel Hack, only 10 are women. These young men follow Chavita's demonstration with their bloodshot eyes—many have barely slept over the course of the weekend hackathon. "Mira todos estos imports" (Look at all these import files), Chavita tells Leo, as he points to the dozens of "import" statements in his Python file. An import statement tells the current file to look at other files that contain previously written code that you can reuse for the task at hand. "No tengo más de cuarenta lineas en cada class" (I don't have more than forty lines in each class), Chavita proudly explains. Leo confirms Chavita's accomplishments with an enthusiastic "Eres un chingón" (You're a badass). Several other young men approve of Leo's designation with a proud nod.

"Cofi"—a hacker from another team that earned this name because of the amount of coffee he consumes—interrupts their final preparations with some *cotorreo*: "Se parece a la Gaviota, ¿no?" (She looks like La Gaviota, no?) he asks as he glances at the the magazine cutout of a topless model his team has posted above their adjacent workspace. ("La Gaviota" [The Seagull] refers to the nickname of President Peña Nieto's wife, who was a telenovela superstar before she became Mexico's first lady.) The young men erupt in laughter, and they take this cue to break out into some noncoding *cotorreo,* a chance for a brief break as the clock winds down.

I take advantage to conduct informal interviews with the hackers. Attempting to get additional information about the motivations behind attendance at this hackathon, I ask Cofi, whom I know is behind on his work for the tech consulting company he works for, "Tú, ¿por qué decidiste venir al hackathon cuando tienes tanto trabajo?" (Why did you decide to come to the hackathon when you have so much work?) "Yo sólo vengo por las chavas," (I only come for the girls), he announces to the rest of the team. We look around the area and the group erupts in laughter. Aside from the magazine cutout of the topless model, there are no "chavas." Indeed, the closest thing to *chavas* some of the hackers have come to see is *Chavita*, the expert programmer who can help them develop a sophisticated program in addition to providing them feedback on their independent creations. "Chava" is a Spanish nickname commonly used for men named Salvador; it's believed the origin is from young children mispronouncing "Salva," short for Salvador, hence "Chava." Interestingly enough, Chavita further infantilizes and feminizes "Chava," arguably to make Salvador's computing prowess more manageable, a way to bring him down, perhaps to "pwn" him.[1]

Indeed, just as one can become a chingón by sharing one's coding acuity, one needs to be careful to not be marked as *too* chingón by showing off. During several occasions at different hackathons, I overheard participants speak of others hackers as "se cree muy chingón" (he thinks he's so badass) or even "ni que fuera tan chingón" (it's not like he's so badass). The mentors at the hackathons were easy targets and frequent victims of these remarks. Hackers who volunteered to be mentors at hackathons and were designated as

1. The hacker and gamer slang term pwn (pronounced "pone") comes from "own," meaning you have gained power or mastery over someone, and is said to have originated when a user mistakenly typed a "p" instead of an "o," the two being adjacent on the QWERTY keyboard.

such by the organizers had usually gained the respect of the community and were considered to be more experienced and able to help participants who might get stuck with coding problems. On a few occasions, I witnessed team members "test" these mentors by calling them over to ask a question about their knowledge of the programming language, more obscure methods of a particular code object, or to help them debug a piece of code. The team members already knew the answer, but by testing the mentor they wanted to show that they knew more than these supposed experts; they wanted to show other team members that they were "más chingones" (more badass).

Analyzed from an outsider's perspective, the expressive forms and cultural performances that make up this hacking space might be recognized as aggressive remarks and sexualized jokes that construct a space where technical masculinity needs to be performed, as Leo, Cofi, Chavita, and "the boys" seemingly just hang out and hack. They might even resemble the *carnales* José Limón conducted research with in southern Texas in the late twentieth century, who were just *llevándosela* and *echando relajo* at a *carne asada*.[2] Limon's carnales exchanged aggressive idioms of sexual violation among themselves, especially using the word *chingar.* "Me chingaron en el jale" (I got screwed at work); "Pos gano Reagan, y ahora si nos van a chingar" (Well, Reagan won, now we're all really going to get screwed); "La vida es una chinga" (Life is being constantly screwed).[3]

Instead of thinking of these individuals as classic machos or homophobes, Limón argues, one needs to consider the sociopolitical context in which they're entangled. "This homosexuality-in-play may also be reversing the sociosexual idiom of *chingar* as practiced by *los chingones* that continually violates the well-being and dignity of these working-class men."[4] That is, these specially marked spaces create moments in which the social world is reversed, in which these working-class *batos* (dudes, guys) become the *chingones*.

While the usage of the word *chingar* predominates in Mexican society, in many cases the word *chingón* refers to an aggressive, astute person. In the context of the hackathon, the frequent circulation of the designation also points to someone who can use their skills to quickly and efficiently solve

2. Limón, 1994. A rough translation here is "hanging out/screwing around at a barbeque." As Limón notes, *carne* (meat) is closely linked to *carnales*, "a kinship term used among brothers or close male friends" (137), used among Mexican-Americans in the Chicano Movement of the 1960s, but also a very common slang term used in Mexico City.

3. Limón, 1994: 132. Translations by José Limón.

4. Limón, 1994: 132.

technical problems. Researching a different setting in Mexico, the Cancún tourist industry, Bianet Castellanos examines the discourse of being chingón as a way for Maya migrants to survive in a new (exploitative) economy with a sense of dignity and agency.[5] When these migrants speak of becoming chingón, they are asserting how they have become adept at maneuvering systems of power and understanding how to use this knowledge to improve their situation; becoming chingón means gaining consciousness of how power works and offering a gendered and racialized critique of the global economy.

Similarly, the male hackers at Angel Hack use the time and space to share code from other projects they have been working on, sometimes from their professional jobs, where results-oriented employers fail to recognize the complexity and beauty of their creations. Cofi tells me, for example, that when he was hired to work for a tech consulting firm in Mexico City, the company gave him a clear message: "Aquí puedes venir en shorts y chanclas, pero vas a trabajar" (You can come here wearing shorts and sandals, but you're going to work). At the hackathon, whether in chanclas or not, they can "be themselves," find solidarity in numbers, and create community with peers who understand and value their code work.

There are many parallels between Limón's batos and these hackers. Instead of flipping *carne* amongst their carnales, here they flip some *código* with their fellow coders. In the same way that Limón argues that his batos prepare and consume their undervalued, low-prestige meat to symbolically take pride in their working-class status, my *coder-batos* craft and share their código and take turns becoming chingón in attempt to call attention to the underappreciation of their trade.[6] But what makes the *code*worlds different from the *carne*worlds? Or are hackathons simply another space where men and normative masculinity come to rule?

It's important to remember that even as new forms of hacking that elevate underrepresented voices and perspectives have emerged, they still exist in the margins. Hackers across Mexico find ways to construct their selves as they cope with the contradictions of coding being valued in some places but not in others. This chapter mobilizes an intersectional analysis of how transnational labor, class, and masculinity come together at hackerspaces to

5. Castellanos, 2011.

6. They are perhaps a new iteration of Gómez-Peñas' (2001) cyber-vatos (spelled with v), who struggled to have their contributions to cyberspace valued in the 1990s.

consider how (mostly male) research participants enact hacker imaginaries and ethics as they strive to become chingones at the hackathon.

[1] Hard Work AND Hard Jefes

While Chavita, Leo, Cofi, and others hack away and consult with each other on their code, they are clearly enacting some of the core tenets from the hacker ethic proposed at the hacker school. By consistently showing up to these spaces, they embody "<9> Get involved"; by working with each other and learning new skills, they become poster children for "<7> Know your tools/communities"; and by joking around and enjoying themselves, they show that although code work can be arduous, they can also "<10> Have fun." These hacker ethics intersect with politics of labor and constructions of masculinity when we consider the different ways one might become a chingón within these hackerspaces.

Castellanos draws on classic writing by Octavio Paz to remind us that the masculine nature of the verb *chingar* "hints at an underlying violence (physical, sexual, and psychological)."[7] Gloria Anzaldúa reaffirms that the verb is gendered and sexualized. She writes, "*Malinali Tenepat,* or *Malintzin,* has become known as *la Chingada*—the fucked one. She has become the bad word that passes a dozen times a day from the lips of Chicanos. Whore, prostitute, the woman who sold out her people to the Spaniards are epithets Chicanos spit out with contempt."[8] For Anzaldúa, the constant usage of the term represents the deep contempt Mexicans and Chicanos (Mexican-Americans) feel toward not only women, but their indigenous selves. "The worst kind of betrayal lies in making us believe that the Indian woman in us is the betrayer. We, *indias y mestizas,* police the Indian in us, brutalize and condemn her. Male culture has done a good job on us."[9]

What both Limón and Castellanos highlight, however, is that the contradictory nature of the term *chingar* lends itself to be used by subjugated groups to create spaces for critiques of power along different dimensions of difference. Limón's carne-flipping batos inverted hierarchies of power on not only white men but also upper-middle-class Mexicans; Castellano's migrant Maya women in Cancún become *chingonas* by showing other women how they can use their observational skills to quickly outsmart people, even

---

7. Castellanos, 2011: 282; Paz, 1985.
8. Anzaldúa, 1987: 22.
9. Anzaldúa, 1987: 22; also quoted in Castellanos, 2011: 279.

formally educated, middle-class anthropologists.[10] Being fearless, inventive, and a problem-solver might earn you the recognition of a chingón/a, but identifying with the term also grants one a strong sense of self in a world where hierarchies of power might try to make you feel like you, or your work, is not valued.[11] But who or what are these young men subjugated by within the space of the hackathon?

Cofi's imagined remodeling of his home workspace, as depicted in this chapter's exvoto, sheds some light on the construction of chingones (and masculinity) at the hackathon.[12] Cofi lived with his father, mother, and three younger siblings in a working-class municipality, Ecatepec de Morelos, northeast of Mexico City. His dad, or *el jefe* as he often referred to him, was a proud car mechanic of over 40 years, who had come to be the owner of the mechanic shop where he was once a worker. El jefe's vision was that Cofi would eventually take over the business and continue in his tradition. That wasn't going to be the case, unless Cofi took over the business with zero mechanic experience, as Cofi had clearly taken to coding instead of mechanic work. There were some similarities between coding and fixing cars, Cofi told me, but el jefe failed to see this.

In fact, many of the hacker ethics Cofi was cultivating at the hacker school and the hackathons seemed to collide, instead of intersect, with el jefe's idea of hard work. The "<3> Doing > Talking" as a hands-on ethic was something only coders could understand; el jefe was unable to recognize sitting in front of a computer all day as the manual, hands-on work central to his labor. All this code work "inside of the computer" meant getting to intimately know the tools available to resolve complex problems, as Cofi mobilized hacker ethics "<7> Know your tools/communities" and "<8> Always be learning." Although el jefe might very well have been undertaking and valuing these very ethics within the mechanic shop, he frequently scolded Cofi for "being in front of his computer all day." "Tal vez si arreglará mi espacio como su taller de mecánica el jefe finalmente podría ver que escribir código también es trabajo duro." (Maybe if I fixed up my workspace like his mechanic shop my dad would be able to see that writing code is also hard work.)

As the artist's exvoto depiction that opens this chapter demonstrates, Cofi imagined that a simple overhaul of his workspace would allow el jefe

10. Castellanos, 2011: 281.

11. Castellanos, 2011: 281.

12. An exvoto is a votive offering to a saint or deity. In the Latin world they usually depict a dangerous incident the offeror survived. In the Mexican context you'll find many related to border crossings, usually offered to the Virgen de Guadalupe.

to see the connections between coding and mechanic work. Cofi's face-tious remarks give us a hint as to why the topless model was positioned above his team's workspace at the hackathon. As a matter of fact, he was the one responsible for posting it there. That he said she looked like Mexico's president's wife reminds us of Limón's batos, who also implicitly incorporate critique of governmentality and politics into their banter. More importantly, Cofi's story brings into perspective how he and other working-class hack-ers in Mexico sometimes struggle to have their labor valued across diverse spaces. Cofi's code work at home was undervalued by his family; his coding on the job was undervalued by his employer; and his code work is at risk of being "used" by state representatives, as the previous chapter analyzes.

These misrecognitions of value across domains of work might have been avoided if el jefe, Cofi's employer, state representatives, and others skeptical of the code work were familiar with what Sareeta Amrute calls "techno-ethics."[13] As Amrute lays out, the first step toward arriving at a techno-ethics is to recognize coding as labor, the kind of work that uses all of a person's mental and physical faculties. "Coding includes hunching over a laptop for hours on end, and can often involve solving problems that coders themselves find uninteresting."[14] "Though it can be fun, and coders get into a flow, it can also be a grind, repetitive and unfulfilling."[15] Within the space of the hack-athon, then, Cofi and other hackers find a place where they can get into the grind, let the code work flow, have some fun, but most importantly, where their code work is valued; the hackathon offers a respite from people and spaces who don't take Amrute's techno-ethics as an assumption.

But what about the hackers whose undervalued technical acumen does not find the respect it deserves at the hackathon? Sasha Costanza-Chock, for example, argues that the hacks that occur within "subaltern design sites" such as the home never receive the same recognition or resources as those marked to be properly technologically innovative within the bounded space of hackerspaces.[16] While the hackerspaces in Mexico offer young men from lower socioeconomic backgrounds the sanitized spaces designed specifi-cally for them to become respectable, middle-class-leaning or white-collar workers, becoming chingones within these spaces also means participating in performances of technical masculinity where they test their knowledge and inadvertently (or not) exclude women and others who do not receive

---

13. Amrute, 2018.
14. Amrute, 2018.
15. Amrute, 2018.
16. Costanza-Chock, 2020: 140.

the same opportunity to build community, learn the tools and skills, and have the "fun" the hacker ethics promote. These questions of who gets to hack, and whose hacking is valued by whom, is further complicated when we consider the transnational dimensions of becoming a hacker and becoming chingón at the hackerspace.

[2] Breaking into OR Breaking out of?

"Tiene cara de llamarse Esteban." (He has an "Esteban" face.) This is how Cofi welcomed the potential ParkInMySpace team member when the hackathon started. Some of them knew his actual name, as he had been a bootcamp instructor at UNAM the previous summer. But after this moment, they just called him Esteban.

After Cofi explained how his managers were okay with him wearing *chanclas* and shorts as long as he actually worked a lot, Esteban added his perspective, "Parece que se trajeron la cultura sin traerse el respeto a los *coders*." (It sounds like they brought the culture without bringing the respect for the coders.) He went on to explain that he had worked for a company in San Francisco where they made it a policy to pay the software developers more than the managers and sales people; this was meant to demonstrate that they valued the work of the coders above all else. Esteban clarified that this was an anomaly and that in truth, the common case was not too far from what Cofi was describing, even in the US. More interested in the practicalities of his employment, another ParkInMySpace member asked, "¿Cómo conseguiste permiso para trabajar en San Francisco?" (How did you get a work permit to work in San Francisco?) "Es Chicano y fue a MIT" (He's Chicano and went to MIT), Leo responded for Esteban. (Leo took almost any chance he could to bring attention to the fact that his friend went to MIT, as it was a place many of the hackers admired and aspired to at least visit one day.) "¿Cómo hablas español entonces?" (How do you speak Spanish then?), another ParkInMySpace member asked. With a growing audience, Esteban proceeded to unpack "Chicano" and his trajectory to MIT.

Esteban grew up in the Los Angeles area, in a Spanish-speaking home, and was the first person in his family to attend college. He chose to attend MIT, where he studied computer science and engineering. Unlike the archetypal "geeks" that fundamental studies on hacker culture introduced us to,[17] Esteban didn't learn to hack on his own or with a "beginner's guide" in his

17. Beltrán, 2022.

parent's basement. This isn't because he wasn't interested in computers per se, but more because his family didn't own a computer until he was close to graduating high school in the early 2000s. He excelled in courses in his high school, and was offered admission to several colleges, but decided to attend MIT because of its "culture," he claimed. "I liked that you could call professors by their first name, and everyone just felt down to earth there," Esteban remembers about his first impressions when he visited campus. On why he decided to study computer science, he tells me, "I thought if I majored in computer science at MIT I would have a guaranteed well-paying job for the rest of my life."

For Esteban, the draw of a college education in general, and learning to code at MIT specifically, was directly connected to the appeal of secure employment and social mobility. This is not an unusual story for MIT students; over 30% of undergraduates are first-generation college students. Esteban's high school was predominantly Latinx, and less than 10% of students attended college. MIT and coding were an invitation to the middle class. As he arrived on the hackathon scene, he took a summer job in Mexico where he showed students at UNAM coding skills, as well as "entrepreneurial" sensibilities; since the six-week boot camp was sponsored by MIT, the organizers trained instructors to instill an MIT "culture," or an entrepreneurial ethos connected to technical competency. As Esteban tells his story, Cofi interrupts, "Yo pensé que eras un hijo de papi" (I thought you were a daddy's boy).

Esteban's story and Cofi's comment raise the contradictions of coding, labor, class, and masculinity. That Cofi thought you had to be an "hijo de papi," or an upper-middle-class young man who relied on his family's status and financial well-being to be successful, reflects the overall disillusionment young men in Mexico experience with the promise of meritocracy. They struggle to find their place between the duped neoliberal subject and the empowered or triumphant coding hero. Esteban, Cofi, and the rest of the coders serve as mentors at other hackathons that promote gender inclusivity, but then they actively create this male-dominated space, complete with posters of topless women.

At the core of these contradictions is the fact that these young men's coding skills are valued in some spaces and seen as emasculating in others, as we saw in the previous section. The connections Esteban makes with Cofi on precisely these issues further elucidate how these dynamics work across national borders. Esteban recounted the following anecdote: "I can never work from home. My dad always interrupts me, and when I tell him

I'm working, he just laughs and says, 'Uy qué trabajo tan duro' (Yeah that's real hard work)." Esteban's father, a migrant from northern Mexico who had worked as a dishwasher and baker for over 40 years, had effectively plugged himself in to the migrant "hard work" ethic that gave him self-esteem and helped carve out a niche for himself and other laborers in the low-wage job market.[18] This politics of "hard work" among migrants in the US forces migrants to place themselves in hierarchies of racialization and deserving-ness among other migrants. As Esteban pointed out, his coding, sitting in front of the computer for hours on end, had not yet made it into the system of values that would recognize this type of labor as "hard work."

Cofi and Esteban thus shared stories about their fathers not being able to value their code work and were able to make these connections themselves. This made sense even if Esteban carried the elite MIT badge of honor, as both hackers came from a working-class background. Cofi was at first simply unfamiliar with the different ways to get into an elite university in the US, only having heard stories of Mexican students from wealthy or well-connected families entering prestigious US universities. Even though they entered the code worlds from different sides of the US/México border, the hackathon in Mexico City had brought them together, and the underlying hacker ethics united them. Here Esteban's first-gen "migrant ethic" might even be said to align with the Mexican hacker ethic. One might infer that Esteban was even (unknowingly) deploying hacker ethic "<1> Give before you get," by coming to Mexico to work with his more underprivileged or under-resourced hacker peers, but his reasons for being at the hackathon were actually more complicated than this.

While Esteban had participated in the hackathon circuit in the US, he found the events and spaces more welcoming in Mexico. "In the States even if I'm from MIT, I'm still a Mexican or Latino coder. In Mexico I'm just a coder. Or just an MIT coder if I pull out that badge." While Esteban was adept at identifying and relating to different hackers across the US/México border in the context of shifting markers of ethnicity, race, and nation, his "escape" to Mexico pointed at an attempt to not only evade the US structures of racialization, but to have his code work valued as "just a coder" before he pulled out the MIT title, if he wanted to take things there.

Scholars have pointed out that hackers from marginalized locations, in particular the Global South, might very well be looking to break into global techno-cultures from which they have been excluded. This is in contrast to the

18. De Genova and Ramos-Zayas, 2003; Gomberg-Muñoz, 2010; Wortham et al., 2009.

"breaking out of" sociotechnical limitations that hacking in the Global North posits.[19] But here Esteban, someone from the "Global North," is effectively trying to *break into* the techno-cultures of the Global South! His ambitions to have his code work valued and build community in Mexico show that cultures of hacking and corresponding constructions of self change the directionality of these practices and efforts when we take into account shifting markers of ethnicity, race, and nation, all inflected by constructions of masculinity.

In the case of the hackers in Mexico, despite their (presumably) ascending authority as "coders," "hackers," or even "chingones" in Mexico, they might still simply be "Mexican coders" once they transport across nationalized borders, as Esteban demonstrates. As Sareeta Amrute shows in the research she conducts with Indian software developers in Berlin, the related "software developer" marker grants workers an entry to the type of cultural capital where one can identify as middle class, but the "Indian" marker prevents workers from becoming anything more than racialized software developers.[20] Race and class intersect across a terrain of transnational labor that values technical expertise yet differentially recognizes and rewards this expertise. Dhaliwal uses the figure of the "cyber-homunculus" to show how the tiny othered science-fictional subjects inside of the machine allow us to peek into the racialized social relations that "buttresses the racial and economic logics of computation itself," continuously devaluing the very labor that drives them.[21] Indeed, these racialized, ethnicized, and always *othered* workers plug into highly coded and particularized lateral spaces, that defy national borders yet rely on infrastructures that mobilize networks and workers to respond efficiently to market conditions, and shift among different systems of codes that enforce ethnic discipline and social cohesion in segregated labor sites.[22] The hackathon can thus be examined as a space where hackers come to perform not only their masculinity but also their willingness to become the coding (and coded) workers of the future.

[3] Other(ed) Hackers

Recent studies of hacking have shifted to consider how markers of difference within hacking communities point to some of the shortcomings of hacking in its idealized form, where ideals of open access, meritocracy, and

19. Nguyen, 2016.
20. Amrute, 2016.
21. Dhaliwal, 2022: 379.
22. Ong, 2006.

transparency rise or fall on the elegance of coding alone. These approaches ground intersectional frameworks to interrogate who gets to count as a hacker and what counts as hacking.[23] Historians have shown not only that woman were central to creating the ENIAC, the first electronic computer, but that what we now call "programming"—transferring skills to automated processes—was initially feminized and considered women's work.[24] Communities have not only fought to highlight these silenced histories but actively aligned different strands of feminist hacking with the goals of specific movements (radical politics in general, antimilitarism, and anticolonialism) and of open-technology cultures.[25] Far from the masculinist practices of dominance or "pwning," Dunbar-Hester, for example, summons bell hooks[26] to point to the platonic love that permeated some of the hacker communities she spent time with; care, trust, honesty, and commitment to community and principle were valued by these communities, as they fused hacking with values of care.[27]

Adding an intersectional analysis to studies of hacking further helps us understand just for whom hacking is valued. Dunbar-Hester also draws from the work of Rayvon Fouché and Ben Chappell to point to the horse-drawn hay rake and the lowrider car as examples of "hacks" by members of racialized populations that had to be defended in the face of mainstream ingenuity; these inventions were rarely portrayed as hacking in a positive, agentic sense.[28] In the case of the lowrider, these hacks were not only devalued but actually criminalized. The hydraulic suspension of lowriders was therefore not only for show but was also a pragmatic modification that allowed cars to ride lower than the California legal limit, but then to be lifted in an encounter with a police officer. These examples highlight how visibility, recognition, and respectability are closely tied to practices of hacking; the increased recognition of hackathon sites as places for "respectable innovation" inadvertently mark other design practices, especially those

23. Moraga and Anzaldúa (1981), Crenshaw (1989), and Oyěwùmí (1997) were among the first scholars to consider how race, gender, class and other markers of difference overlap and intersect.

24. Light, 1999.

25. Dunbar-Hester, 2020.

26. hooks, 2001.

27. Dunbar-Hester, 2020: 179. Gil (2022) also argues that Fab Lab Livre's pink legacies in São Paulo challenge US- and European-centered innovation paradigms, allowing for a multiplicity of practices that enable participants to be productive *and* caring.

28. Chappell, 2001; Dunbar-Hester, 2020; Fouché, 2003.

attributed to marginalized populations as "not innovative" or in the worst case, criminal.

Moreover, the emergence of "life hacking"[29] as a mode of creating clever and elegant solutions many times expunges members of the disability community who were the "original lifehackers," consistently developing creative ways to make their worlds more accessible.[30] When hackathons do address issues affecting the disability community, they usually do so with a hacker solutionism that looks for saviors and protagonists, instead of talking with disabled persons themselves. Melanie Yergeau calls the majority of hackathons the "hipster version of telethons," where hacking as passing, fixing, and retrofitting focuses on the normalization of bodies by emphasizing fixing, curing, and rehabilitating people.[31] Yergeau calls for "criptastic hacking," a version of hacking that challenges forced normalization and moves from body-tweaking and patches designed by non-disabled people to a collective, disability-led movement.[32]

What these intersectional analyses and hackathons organized by marginalized communities themselves offer then, is the potential for solidarity and critique. Communities looking to join the hacker worlds are often times framed along their corresponding markers of difference along dimensions of race, gender, and as these scholars and activists call out, able-bodiedness. When organizers from these communities are at the forefront of participating and promoting these events, they intervene in the techno-solutionism that many hackathon initiatives attempt to advance. When techno-entrepreneurial ambitions, along with efforts to become empowered coders, merge with social justice causes, the resulting outcomes can be transformative or liberating only if the resulting hacker imaginaries and ethics take into account the nuanced intersections that result when hierarchies of computing expertise are re-established and re-shuffled.

29. See Reagle, 2019.

30. Jackson, 2018. Boellstorff (2019) further finds that practices by the disability community in digital social worlds challenge the ableist paradigms that structure both the digital social worlds and conceptions of labor, and reframe disability as a form of social action irreducible to limitation or lack.

31. Yergeau, 2014.

32. There are representative examples of projects led by the disability community (https:// thewalrus.ca/hacking-diabetes/) and hacker events that are careful about the "engineer's trap" (https://hackaday.com/tag/disability/). To be clear, many people come to the hacker worlds because they belong to the disability community.

[4] Hacking Imaginaries, Origins, and Intersections

I opened this chapter by asking whether hackathons were just one more space where men and normative masculinity came to rule. To compare the *carne*worlds to the *code*worlds means to consider how the cultures of hacking in Mexico and corresponding formulations of self are negotiated when markers of ethnicity, race, and nationality intersect with values of coding and labor, always inflected by constructions of masculinity. By taking a closer look at the subjectivities that personalities such as Cofi and Esteban bring to the hackathon, it becomes clear that the hacker ethics intersect with, align with, and sometimes challenge ethics formulated from the life experiences of young men who have to face the contradictions of their coding skills being valued in some spaces and undervalued in others.

Intersectional analyses of the hacker imaginaries and subjectivities that people bring to hackerspaces in other contexts further illuminate that the organizers and participants at the events, whether they intend to or not, are always enmeshed in the multiple ways that work and coding get co-produced with race, class, gender, and even able-bodiedness. These markers of difference are often sequestered into the "personal" layer of the ethno-stack, especially by corporate or state entities who have a stake in viewing these systemic issues as merely affecting individuals so that they can continue "business as usual." In this chapter, though, I've teased out how this "personal" layer of the ethno-stack is coupled or de-coupled, brought together or kept separate from the other layers—the interpersonal and especially the sociopolitical—by employers, family dynamics, and the hackers themselves. By adding a transnational lens to studies of hacking, in this case across the US/México border, I've argued that carefully considering these life stories and subjectivities effectively inverts the breaking *into* or *out of* directionality of hacking practices. Coders from the Global South are usually aspiring to break into global techno-cultures from which they have been excluded, but Esteban's origin story shows how his desire to *break into* the techno-cultures of the Global South can only be understood by unpacking shifting labels attached to the generic "coder." These markers come from processes of racialization and class that, as scholars who use intersectional analyses of hacking have shown, leads to some people having their hacking celebrated and others having their hacking criminalized.

Hacker imaginaries in Mexico, as dictated by the hacker ethics formulated at the hacker school, may very well resemble some of the hacker practices from other places in the world, especially those emanating from US culture. But by looking at how Cofi finds similarities with Esteban's migrant "hard work" ethics, we understand how some of the imaginaries are awakened, revitalized, and organized not only by ideas from somewhere "out there" but by people moving across the US/México border and thinking about the differences across the techno-Borderlands. These movements and interactions are central to understanding how these imaginaries become practices and communities, and how evolving subjectivities are looped back into corresponding hacker ethics and subjectivities.

This intersectional analysis aimed at uncovering what it means to become a chingón at the hackathon adds ethnographic texture to the study of how infrastructures of technology and the knowledge economy intersect with mappings of value onto labor as well as constructions of masculinity. If the previous chapter demonstrates how Mexican hackers attempted to distinguish themselves from the "men in suits" (or those in power) who misprize their coding, this chapter shows how young men further construct masculinist approaches to coding in order to counter the undervaluing (and even emasculation) of their work by their own families and employers. In Chapter 4 we'll see how the all-women's hackathon is organized explicitly in opposition to these approaches. The next chapter considers how the subjectivities underlying the code work is welcomed, or not, in domains of life outside of hackerspaces.

[3] Code Work
across Domains

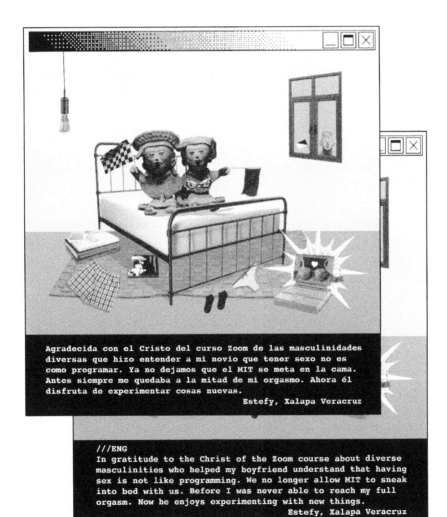

Agradecida con el Cristo del curso Zoom de las masculinidades diversas que hizo entender a mi novio que tener sexo no es como programar. Ya no dejamos que el MIT se meta en la cama. Antes siempre me quedaba a la mitad de mi orgasmo. Ahora él disfruta de experimentar cosas nuevas.

Estefy, Xalapa Veracruz

///ENG
In gratitude to the Christ of the Zoom course about diverse masculinities who helped my boyfriend understand that having sex is not like programming. We no longer allow MIT to sneak into bed with us. Before I was never able to reach my full orgasm. Now he enjoys experimenting with new things.

Estefy, Xalapa Veracruz

[0] Stories in the Time of Hacking

If one could hack away in the code worlds, why not hack the world around us? As I moved across different spaces with my research participants, from the hackathon to their university spaces to the hacker school, I wanted to know how much of their hacker selves carried over to everyday spaces that seemed to be removed from the worlds of hacking. In her ethnography of Wall Street, Karen Ho studies the construction of financial markets and busting of market bubbles.[1] She focuses on the daily practices, the values and dispositions that come to define Wall Street's cultural practices. She argues that investment bankers see themselves as incarnations of the market—they become one with its rhythms and movements. I wondered, were there similar dynamics at play for code workers? Did they become one with logics that drive the code worlds, with the rhythms and movements of the worlds of computing?

Historians and anthropologists of computing have provided the foundations to understand that what's in the code is always influenced by what's outside of the code, and vice versa. That is, to think with computers really means to think about social relations, to explore the abstractions across the layers of the computing stack as always being co-produced with the ideas and stories that shape its use.[2] The very style and approach one brings to coding, the materiality of the code as it appears across the stack's layers of abstraction, is influenced by the dispositions carefully cultivated by coders outside of the code worlds, and vice versa.[3] Cognitivist theories on analogy-making mostly treat metaphors as motivated by lack or deficiency in one domain. That is, they find that subjects deploy a cognitivist need to understand things across domains by using metaphors from one domain to understand something in the other. But, as Wendy Chun asserts, "Software as metaphor for metaphor troubles the usual functioning of metaphor, that is, the clarification of an unknown concept through a known one. For, if software illuminates an unknown, it does so through an unknowable

1. Ho, 2009.
2. Abbate and Dick, 2022: 2–3.
3. Coleman, 2013; Fuller, 2008.

(software)."[4] To get to this idea that there's always a back and forth, between the code and social life, and inspired by software's appeal to uncover an "unknown using an unknown," to trace how the code work might travel in unexpected ways to surprise domains, I had my research participants reflect on these crossings themselves. I approached them with an open-ended question: "What other aspects of your life do you feel are most influenced by your identity/role as a software developer? Can you think of examples that illustrate this?"[5]

In this chapter I demonstrate how hackers bring technical idioms from the code worlds to all sorts of (non-technical) aspects of their lives: their approach to gardening and cooking, the way they view dating and romantic relationships, and even their predicaments as migrants or transnational subjects. Their everyday stories, in response to my prompt, reveal that their commitments to the hacker ethic, their strong sense of self and their approach to self-cultivation, their reliance on their hacker communities, and their predilection toward careful planning and efficiency were carried over to other domains of life—and in fact, they believed that the world might be better if we all hacked the world around us in similar ways. I demonstrate how elements that hacker-entrepreneurs associate with coding (speed, iterative thinking) and the entrepreneurial worlds (contracts, ephemerality) infiltrate their personal and social lives, and even their dating and sex lives, whether they planned for it or not. While some of the resulting stories might come off as entertaining for some, I end on a more serious tone with a story that addresses healthcare inequities as they relate to classed and racialized differences in the US. I conclude that generic tropes such as the "compulsive programmer" are inadequate to fully comprehend the ways a form of "coder's paranoia" intersects with a migrant paranoia that affects hackers navigating the US/México techno-Borderlands more profoundly than the average unmarked hacker.

[1] Problem-Solving across Domains

My seemingly straightforward question about how their identities as coders connected to other parts of their lives was understood by respondents in varied ways and thus received a diverse set of responses. One thread of

---

4. Chun et al., 2022.

5. For analysis on the politics of asking and interviewing in ethnographic projects, see Briggs, 1986, 2007.

answers fell along the line of "no separation." These developers wanted me to understand that there was no distinction between their roles as software developers and the other aspects of their lives. They carried their identities with them all of the time even as they separated themselves physically from their computers and even when they weren't specifically in the act of programming. Cofi, for example, told me, "It's not like I can suddenly turn it off or on. I'm always a coder and I'm always thinking like a coder. . . . When I shower, when I eat, even when I sleep. I dream of code." Cofi wanted me to understand that his coder self was never compromised and that it wasn't something that he could simply leave aside. The reference to his everyday activities and especially to sleeping made it clear that it wasn't even something that was particularly conscious.

The reference to sleeping came up several times in responses. Hackers stated that thinking about a solution to a particular problem accompanied them throughout the day, even into their sleep, and that sometimes the solution came in their dreams. Leo added that even when he had already written a particular function or snippet of code given the constraints, and he thought he had let it go, that he would dream about a more efficient or elegant solution and quickly change the function in the morning when he awoke. While some readers may be able to identify with the preoccupations with work or other aspects of their lives that sometimes carry into sleep and into dreams, the responses seemed to confirm some of the imaginaries of the "compulsive programmers" constructed from early studies of geeks and hackers.[6] While this line of responses reveals the dedication that many hackers bring to their craft, other responses highlight how their engagement with code also influenced other social practices that might seem removed from the code worlds.

In another line of responses, the hackers I interviewed referenced projects where they relied on community in some form. Luis explained to me how he approached repairing his car. An enthusiast of the classic Volkswagen Beetle, or the *vocho/vochito* (beetle/little beetle) as it's called in Mexico, Luis took care of a 1979 baby blue vochito that he had inherited from his grandfather and that he kept in pristine condition. The repairs were always anxiety producing, he told me, giving me examples of when what he perceived to be unscrupulous auto mechanics would replace his original parts with generic ones, or when they would overcharge him for surprise services. Instead, he said he decided to try the repairs himself. In his latest repair endeavor, he

6. Weizenbaum, 1976; Turkle, 1984.

had ventured to attempt to fix a leak in the gearbox in his beloved vochito. By learning as much as he could from online tutorials and old manuals, he had assessed that he could possibly get away with just using a sealing kit and reassembling the gearbox. He would attempt the repair, and then, if he got stuck, reach out to his beloved hacker community.

"Aunque a algunos de mis cuates hackers no les gusten los coches como a mí, compartimos la misma maña para moverle y resolver cualquier problema técnico. Yo sé que me apoyan en cualquier proyecto así" (Even if some of my hacker buddies are not into cars in the same way I am, we share the same knack for tinkering and solving any technical problem. They have my back in any project like this), Luis told me. He said he trusted his community more than he trusted the money he could save to hire an auto mechanic expert to help him. His approach and quotes show that he was adhering strictly to hacker ethic #1, "Give before you get," #8, "Always be learning," and #7, "Know your tools/communities."

Luis' valorization of self-learning and technical aptitude, backed up by the confidence in his hacker community reflects similar research findings with hacker collectives. In 2016, I conducted an ethnographic project where I surveyed GitHub software developers about their use of open-source licenses. Approximately 75% of the 35 software developers I conducted open-ended interviews with responded that they used the MIT License for their code contributions to the platform. Some said that they chose this specific license over others (e.g., the GNU General Public Licence) because it was simple to use and understand, or because it was the least restrictive or most permissive license. The majority of respondents who chose this license, however, mentioned something about the importance of community formation in their decisions. Put simply, many chose the MIT License simply because peers they trusted also used the license. Few had actual concrete knowledge about the workings of this license in relation to others; it was enough to know that their respected peers were using the MIT License. In this sense, Luis' response to my prompt, similar to other responses, expressed a deep confidence and commitment to belonging to a community of hackers whom they trusted. In the GitHub case, the trust was enough to decide where your work (in the form of code contributions) would end up and who could use it; in the Luis case it was enough to jump into a new project trusting that your community would "have your back" should things go awry.

Moreover, Luis's comments also connect with the ethics of self-exploration found in other hacker communities. After conducting research with open-source software developers, Coleman found that their cultivation

of technical prowess and expertise is built on the idea of an asymptotic process of self-cultivation.[7] Driven by principles of US-centric meritocracy,
hackers are awarded respect and corresponding privileges in the community
based on their individual technological contributions.[8] Ultimately, a developer must provide the community of hackers with resources such as documentation and, of course, the fruits of their labor: source code. Guided by
principles of individualism and self-mastery, hackers can ask but not ask
too much from community members, as they might receive the dreaded
RTFM (Read the Fucking Manual) response.[9] While the RTFM response
is more aggressive than what might be envisioned in the "Give before you
get" hacker ethic espoused at Luis' hacker school, his approach to the repair
of his vochito adhered to this general outlook—he would get as far as he
could on his own to reassemble or replace his gearbox, and then reach out
to his community when he got stuck. His peers would then presumably
respect his efforts and decide to help him because he had already given it
his best effort. "Give before you get."

A third line of responses to my prompt about how elements of *code work*
showed up across domains continued to revolve around the way in which
coders approached particular activities. Respondents mentioned everyday activities such as cooking and gardening as well as recreational social
practices such as camping to highlight how they brought the same organizational and planning approaches, based on iterative thinking, to these
domains outside of the coding realms. TecnoChica, or T.C., first responded
with a response more aligned with the first line of responses, that there was
no separation between their hacker self and their selves in other domains of
life; at first they didn't provide a concrete example. A few days later, upon
further reflection, they contacted me to let me know that they had some
examples to backup their first response. Their first example was cooking.
They explained to me that before they even started cooking, they knew
exactly what they were going to do. They had the recipe and a plan, they
knew where the ingredients were, and they even knew how long it was going
to take. This was quite different from their partner's cooking style, who got
into the kitchen without a plan necessarily and frequently had to run out (or
send T.C.) to the corner store mid-preparation because they were missing

7. Coleman, 2013.

8. A perfect example of this from the GitHub world is the contribution of visualization graphs
on users' profiles that show how much code users have contributed across their coder lifespans.

9. Coleman, 2013: 111.

an ingredient. These self-reflections were based on comparisons with their partner's approach to activities, as T.C. elaborated in another example:

"El otro día mi pareja se sintió con ansias. Esto, porque el jardín necesitaba una manita para mejorarlo y pensaba que nunca lo haría y que lo pospondría por mucho tiempo. Pero en realidad yo lo estuve pensando en mi cabeza durante días y lo tenía todo visualizado. Después sucedió que hice todo como en dos horas, rápido y eficiente. Franz cuando vio, se ha quedado super impresionada." (The other day my partner was feeling anxious. The garden needed a little hand and they thought I was never going to do it and that I would put it off for a long time. But I was actually thinking about what I was going to do for days. It was all in my head. Then one day I did it all in like 2 hours, fast and efficient. Franz saw and was super impressed.)

T.C.'s use of the words "rápido y eficiente" (fast and efficient) made it clear that their approach to everyday activities was carefully planned and streamlined, always with an eye toward saving labor. In fact, these qualities are representative of what coders might refer to as beautiful code. And they certainly identify the code a good coder writes: fast, efficient, well organized, and always without unnecessary lag or repetition.

Lotar added another activity along this line of reasoning when he mentioned a camping trip he participated in on a recent trip to California:

"Creo que es porque nunca he acampado pero investigué un montón antes acerca de esta actividad. Me gusta tener la mayor cantidad de información posible para el viaje. Así es como le entro a cualquier nuevo proyecto. Hago un chingo de investigación y me gusta tener opciones y herramientas disponibles antes de entrarle a cualquier cosa. Para cuando llegue el momento hacerlo de la manera más eficiente posible."

(I think it's because I had never been camping before, but I did a lot of research about the activity. I wanted to have the most possible information for the trip. That's how I approach any new project. I do a shit load of research and I like to have options and the best possible tools available before I jump into anything new, so that when the time comes I can do it as efficiently as possible.)

Lotar explicitly references the first part of hacker ethic <7>, "Know your tools/communities," and by including the keyword eficiente (efficient) in his response, he further confirms the connection to self-learning and careful preparation that many coders see themselves possessing and bringing into other domains of their social lives. That both Lotar and T.C. associated these with coder identities reveals what they consider to be the subjectivities that align with coder identities, or at least with the practices of a good coder. Moreover, they both employ a sort of iterative thinking when they plan their activities. They ask themselves how to approach their tasks so that they don't have to come back to get another ingredient or another tool. They think about the assignment and iterate through the possibilities in their heads before they actually execute the job, in the same way they might write the instructions inside of a loop so that they write the minimum lines of code possible.

Most of the responses from my research participants showed that they saw these cross-domain traversals, from the code to the social, as positive. They agreed that the ethics of hacking and logics from the code worlds were useful approaches for other domains of life, and in fact, the world might be better if we all hacked not only the code worlds, but the world around us. In the next section I explore the flip side of this. What happens when hacking the world goes wrong?

[2] In the Mood for Love in (and out of) the Code Worlds

If my research participants brought their hacking sensibilities from the code worlds to other domains of life, then undoubtedly they also brought them to the love worlds. But how do the core tenets of computer science, based on speed, efficiency, and adding (or eliminating) redundancy when necessary, transfer to people's dating and sex lives? Fortunately some of my research participants were happy to help me understand these traversals when they reflected on my prompt. And as we might expect, some of their stories can be properly filed in the "when hacking goes wrong" category.

One respondent who had substantial experience with attempting to make the laws of the code worlds function in the love worlds was Hiro. Hiro understood my question quite well and it made him come alive with stories he felt demonstrated how his practices as a full-stack developer transferred to other areas of his life.

One set of stories Hiro told me were about how he attempted to hack the dating scene. During a period when he was in San Francisco working for a company on the next generation of wireless chargers, he was being paid well, but in San Francisco it was just enough to pay rent and eat. Dating, he reminded me, was expensive! Even if both parties agreed to pay for their own meals and entertainment, dating still required that Hiro spend more than he would have budgeted in order to go out. So he came up with a hack. He bought himself an annual membership to the San Francisco Exploratorium. For a $60 fee, he got unlimited visits, and the annual membership allowed him to bring a guest for free. With the regular entrance fee of $10, he only had to go on more than 3 dates to make it a good value. But the membership benefits also included two free drinks. "Because in this country they are so liberal," he explained to me, "my dates end up offering to pay for both drinks, since I paid for the entrance. I go get the drinks, they don't see that they were actually included, and they end up Venmo-ing for the drinks. I've actually made money off of the membership!" Hiro recounted his dating hack with much enthusiasm, letting me know that this actually encouraged him to go on dates every week. He proceeded to provide me with extended accounts of acquaintances whom the dating scene had left broke(n), if not emotionally at least financially. This Exploratorium hack allowed Hiro to hang in there, but with his anecdote he also alluded to the sort of ethos of "gaming the system" that he brought to the dating experience, which only became more evident (and more exaggerated) as he told me more stories.

Sometimes "hacking the dating scene" went wrong. Because Hiro went on dates frequently to make the annual membership a good value, and because there were times when these series of dates overlapped with various (potential) partners, Hiro made it a case to find partners with the same name. He told me the story of the two Marias. He was having coffee with Maria #1 when Maria #2 texted him, looking to make plans for later that evening. Hiro stepped out of the coffee shop, and texted Maria that he would call her later to make plans since he was currently on a business lunch. Of course, in his hurried attempt to maintain the possibilities open with both Marias, Hiro accidentally texted Maria #1, who was inside of the coffee shop. As he walked back in, she greeted him with a perplexed look and asked "Why did you just text me this?" as she held up her phone in dismay. Maria #1 didn't finish her $7 iced coffee mint mojito. Hiro asked for his to go. It was their last date. Hiro explained to me that it was a moment when "adding redundancy to the system had gone wrong." While he was being facetious in his recounting of the anecdote, and I think that dating the two Marias simultaneously was more coincidence than strategic maneuvering, the fact that he gave this story as part of his response to my prompt and that he referenced redundancy was quite revealing. In computer systems design integrating redundancy can mean duplicating critical components or functions of a system in case one fails to make the system more reliable. In this case, dating two persons at the same time provided Hiro with some sort of security that at least one would maybe work out. That they both had the same name seemed to provide security that he wouldn't screw things up by mixing up their names during a date. It didn't quite work out that way, as we saw, and for those wondering—it also didn't work out with Maria #2.[10] Most of the anecdotes Hiro conveyed to me were actually about mishaps in the dating world. In fact, that he never heard again from Maria #1 wasn't the first time he had been "ghosted."

Returning to his Exploratorium hack, I asked Hiro if he didn't get bored by going so many times to the same place. "It was boring since the first time I went," he responded. "The less physics and math you know, the more magical that place seems." Apparently Hiro wasn't enthralled by the hundreds of exhibits inside the galleries that promised to provoke "joyful exploration" that would allow guests' "curiosity to roam free." Hiro seemed to bring his "how things work" and systematic approach to explaining the behind the scenes "magic" the Exploratorium sold. Some of Hiro's dates must have also felt his boredom or weren't excited by his approach, which would explain why some of them "ghosted" him during the date. "One told me she was going to the bathroom and never came back," Hiro laughed. Fortunately for him, this disillusioned date still Venmo-ed him for the drinks.

Having dates go wrong might not be particular to participants of the code worlds. Misadventures and faux pas in the dating world are the stuff that makes television series and interesting stories to share with friends across many social and professional circles. But this lack of "magic" in dates and romantic relationships crept up in interesting ways in my respondent's stories. Rodo's response to my prompt, about how his life as a software developer influenced other aspects of his social life, started with, "Well, my girlfriend once told me she didn't want MIT in bed with us."[11] Rodo was a student in the MIT summer coding bootcamp I helped teach in Xalapa in the summer of 2014. While I initially visualized his response quite literally, that Rodo was perhaps so attached to his laptop that he wanted to keep working on the coding assignments from the MIT course even while he was in his bed, his story was more complicated than this.

For Rodo's girlfriend, Estefy, having MIT in bed with them referenced the way in which Rodo was approaching their sexual life. "It wasn't because I'm always thinking about the code. She was really annoyed with my *mala chamba* [doing a bad job], with the way I was translating the fastness and

10. Neither Maria #1 nor Maria #2 were available for comment.
11. Quote verbatim, in English.

efficiency of a good algorithm into my performance in bed,"[12] Rodo started to reflect. He believed Estefy wanted him to slow down, take his time, and concentrate on the task at hand. Rodo had been discussing the MIT course with Estefy, a visual arts student, throughout the summer. He had mentioned our lessons on the core principles of writing good algorithms—their simple elegance defined by their speed and efficient execution. So Estefy was associating this approach to their sex life, to which she felt Rodo was bringing an unwelcomed "MIT approach." On the one hand, Rodo told me that sometimes she would get upset because he seemed to be somewhere else during their romantic couplings. He admitted that he was many times indeed thinking about how to improve his code, like the "compulsive programmers" discussed in the previous section. Moreover, Estefy yearned for a sexual experience that wasn't defined by the tempo dictated by the world of efficient coding, with MIT representing the apex of these practices. That she didn't want MIT in bed with them seemed to be a warning to Rodo to leave the hacking in the code worlds, or at least at the door before he entered the bedroom. Interestingly enough, Estefy subsequently worked on a startup project named SugarNut, a vibrator that connected the user to anonymous users who could control the intensity and patterns of the vibrations over wifi.

Rodo and Estefy's story is reminiscent of the ways that the cultures of hacking have interrupted the sexual lives of couples throughout history. Kristen Haring, for example, details a cultural history of ham radio that is considered to be one of the early influences of cultures of hacking in the US.[13] The DIY cultural ethos that permeated these communities was a direct result of post-WWII narratives that encouraged parents to allow young men to tinker with technical components as a means to promote engineering and scientific education, which could lead to national prosperity. Haring argues that in order to escape de-industrialization and the corresponding feminization of work, men constructed shacks or garages as separate spaces within the normative suburban home life, where women and children would not interrupt them and where they could connect with other men to reclaim their hands-on, brawny masculinity.[14] But this culture's appeal to technical fraternity and technical interactivity as a means to express masculinity broke the norm of midcentury, middle-class standards. Previously, their masculinity was performed in the domestic context through their relationship with

12. Quote verbatim, in Spanglish.
13. Haring, 2006.
14. Gelber, 1997; Haring, 2006.

women; a masculinity now based on other men and machines disrupted this relationship and carried sexual insinuations.[15] In one story, Haring's respondent (Ann) tells about her "romantic introduction to ham radio." Ann drives to a scenic lookout spot with her partner Gordon, and just as she begins to feel "in the spirit of things," out comes a microphone and switches, and the intimate moment begins with the distant voice of a male stranger.[16] Suffice to say that she quickly fell out of the mood, and the romantic escape was ruined. The ham radio had escaped the confines of the domestic garage and was now infiltrating their moments of passion, even as she they tried to carve out a space just for themselves. These sentiments were echoed by other women whose partners seemed more interested in their electronic toys than them, with some even joking that they were looking for another husband for "upstairs use."[17]

Rodo's story continues in this history of the practices of hacking transgressing into spaces where they are not welcome. Rodo didn't bring the microphone and switches of the ham days into bed with him, but he did bring an approach that bothered Estefy and that she associated with the world of coding represented by MIT and the coding bootcamp. Similarly, Cofi revealed to me that his partner had even named his laptop to express her disapproval of his lack of attention to her. "Me dice cosas como 'Vas a cenar conmigo o ya tienes planes con Mildred'" (She'll say things like, 'Are you going to be able to have dinner with me or do you have plans with Mildred'), he told me. Mildred was not a name he used but one that his partner had assigned to his laptop, fully personifying the machine as an entity that had an important enough presence to intervene in their plans. The code worlds, the named machine that carried them, and the institutional leviathan that represented them had come to interfere in my respondents' quest for love and romance in ways that they had not expected.

To be fair to my respondents, I have to mention that some did listen to their partner's complaints and that there were attempts to rectify their shortcomings. On one of my return trips to Mexico City, about five years after my initial fieldwork and interviews, I visited Rodo, and I was happy to hear that he was still with Estefy. He was quite excited to tell me about one of his recent attempts to limit the degree to which he transferred the logics

15. Haring, 2006: 126.
16. Haring, 2006: 127.
17. Haring, 2006: 127.

of the code worlds to the magic of the bedroom. He had signed up for an "alternative masculinities" Zoom workshop. These types of extracurricular and self-help workshops had proliferated and become increasing popular in the COVID-19 era, with people becoming increasingly bored in their homes and experts from diverse domains increasing access to their services using social media and Zoom.

Rodo hyped me up on the course, telling me that it was not only super informational but that for him it had taught him how to be a better lover. Intrigued, I signed up for the Zoom course and connected with the instructor, a self-proclaimed alternative masculinities guru, as well as with Spanish speakers across Latin America for two days. As the guru led us on a journey to unpack masculinity, we learned about the myths of penis size, the importance of taking focus away from penetration in order to fully enjoy sex, self-exploration and practicing orgasm control with masturbation to concentrate on pleasuring your partner, and practiced "being there" by closing our eyes and eating our favorite fruit slowly. I chose a mango, especially because I had privileged access to fresher, juicier ones in Mexico than I did in the US. Eventually we introduced ourselves to each other, our names, professions/backgrounds, and what had inspired us to take the course. I was somewhat disillusioned when I heard that none of the other participants was a software developer. In a perfect conclusion to this section, I would have told you that there were many engineers and computer scientists working collaboratively on deconstructing their masculinist, technical approaches to the world. The truth is that the majority of the participants shared a similar profile: artists, academics, nature-loving types; most had long hair and scruffy beards; many lit up spliffs as we started the course. Now every time I'm eating a mango, I think of Rodo's stories, as well as the stories of Estefy, Cofi, T.C., Lotar, and Hiro, as they searched for love, in and out of the code worlds.

[3] 0s and 1s: Between Migrant and Coder Paranoias

As I listened to my respondents' answers to my prompt, I was able to connect to many of their stories. While the hacker ethic and the logics of the code worlds seemed to be getting in the way of the love worlds for some of my research participants, Armios's story further elaborated how this particular approach or way of thinking complicated his relationship. His transnational story brought into focus how elements of racialization and class distinctions also made their way into these cross-domain traversals and exacerbated the "coder's paranoia" some hackers describe.

When I gave him the prompt, he also gave me general responses along the lines of "there is no separation, I try to hack everything," but then gained more trust with me to eventually share a more personal story. Armios was born in the US to Mexican migrant parents and held both US and Mexican passports. The anecdote he shared was about the time he and his partner went for their first infertility treatment appointment in the US. Since he and his partner had been trying to have a baby for more than a year with no success, they now officially qualified to see a specialist, as the state mandated that insurance cover corresponding diagnosis and treatment. When the specialist doctor received them in her office, she asked a series of questions about their experience, how long they had been trying to become pregnant, and so on, and then asked his partner, "Will you be here for more than a year?" His partner, who was in the process of learning English, answered, "No." Armios quickly jumped in and corrected her, "Yes, we will be here for more than year; we live here now." When he later asked his partner why she had answered with a "no," she told him that she understood, "Have you been here for more than year." What seemed to her like a simple mistake was a much more serious misunderstanding for Armios, which provoked him to immediately jump in to correct her. "I didn't want them to think that we were here just taking their resources. Now I'm always super nervous that she's going to flip a yes for a no and they're going to think we're just trying to take advantage of the system, to get services for free, or to confuse us for Mexicans who come to the US just to have babies."[18]

Armios's anecdote and his comments need to be unpacked along various dimensions. What did this story have to do with the code worlds? Why had Armios chosen to share this as an example of crossings from the software development worlds to other domains of life? He went on to explain his experiences as a coder. Like other software programmers, he spent much of his time thinking about how to make algorithms more efficient, thinking about the patterns in the code that would lead to simpler, more elegant functions—that beautiful code many hackers are after. This sort of obsession sometimes led to dreaming of solutions, quite literally. The code made its way into his sleep and dream worlds like it happened to Cofi, Leo, and many other respondents. Another aspect of this coding obsession that he especially recalled quite vividly from his days of undergraduate studies was the fear of something in his code changing without his knowledge. Rumors circulated within his university's computer science department of the malicious

---

18. Quote verbatim, in English.

colleague who would flip a 0 to a 1, or a Boolean from True to False, some-where deep in one of the functions, so that his entire code's logic would be inverted. At worst your course project might be ruined, and at best you would need to spend countless hours debugging the code to find the error.

When Armios told me this, I immediately understood this coder's para-noia and recalled from my undergraduate days studying computer science, when I was fully immersed in the code worlds more than at any other point in my life, similar rumors of the malicious roommate who would flip that 0 to a 1 in your code. At MIT, Course 6 (computer science major) was con-sidered one of the more challenging majors, perhaps not strictly because of the difficulty of the material, but because of the time demands of the classes and their coding projects. Course 6 students spent countless hours diligently searching for patterns and possible abstractions to subsequently write care-ful code connected by intricately related system interfaces. The thought that these meticulous arrangements could fall apart with the simple substitution of a 0 with a 1 led to a type of coder's paranoia, a desire to protect your code.[19]

For Armios, then, the thought that his partner might substitute a yes for a no was similar to that fear of replacing a 0 with a 1. In both cases, there is a risk of *something* falling apart. In the code, the project or system potentially stops functioning or breaks down. In the infertility appointment, there is something much more important at stake, and this was clear by the way Armios told his story. The fact that his analogic thinking led him to connect these two potential "failures" pointed to the facility with which he con-nected two very different formally describable systems. As other authors have noted, programmers possess an affinity for *connecting* which manifests as making associations across formally describable systems, such as binary code, language, and mathematics.[20] Armios recognized that one small error or misunderstanding might lead to a complete misunderstanding of who he and his partner were and why they were there. But this also has to do with elements beyond the coding logics themselves. Here the risk was that he and his partner would be thought of as subjects who were "trying to take advantage of the system." After all, Armios and his partner had followed the rules. They waited the year, approached the right doctors, and were now in

19. For many contemporary software developers who use higher-order programming lan-guages, version control systems, password-protected computers, and cloud storage, the malicious coder might not register as a realistic fear. At places like MIT, where students are taught to use minimalist languages and text editors to teach the fundamentals of programming, the prospect of the prankster roommate is more real.

20. Coleman, 2013; Gaboulry 2013.

line to receive the infertility treatment that the state had mandated would have to be covered by his health insurance.

The fact that his partner was speaking accented English and that they were racialized as Latina/os within the US cannot be separated from the paranoia Armios was feeling. When he says that he doesn't want to be confused for "Mexicans who come to the US just to have babies," he's trying to disassociate himself from the persistent rhetoric of "anchor babies." Leo Chavez examines the use of this pejorative term in political and social contexts, and especially how it is used in public discourse to vilify migrants, leading to US-born citizens (especially of Mexican descent) experiencing trauma, generation after generation.[21] These narratives that claim Mexican mothers come to the US to have babies just to better their lives by taking resources from the state fall in line with recurring discourses that frame Latinxs as a "threat" to the nation, deploying metaphors of contagion, invasion, and disease.[22] The perplexing thing is that sometimes these discourses frame this collective as an economic liability who take jobs, resources, and benefits from "regular" Americans, but other times they are presented (especially by mainstream media) as grounded in distinctive American values of hard work and family, as orderly, sanitized, respectable, middle-class, employed citizens.[23] This dizzying effect of how and why and to whom one should identify as "Latinx" (US American) or "Mexican," a form of racialized migrant paranoia, led Armios to make the analogic connection to a 0 being substituted for a 1, a type of coder's paranoia.

To make his migrant's paranoia more severe, Armios and his partner were also going through this difficult process during a Trump presidency, replete with xenophobic and racist comments but also amendments to naturalization laws that made the "public charge" rules more strict; citizens-to-be would possibly have their applications rejected if they were deemed to be possible burdens on the state in the future. While these laws didn't actually apply to Armios and his partner—she was going through the naturalization process, so the fear that they might be racialized in this way and confused for "undesirable national subjects" contributed to his paranoia—confusing that one "yes" for a "no" put the entire interaction and relationship with their doctor at risk. His coder's paranoia, and the way he transferred these feelings from the code worlds to his personal life, was affected by this particular

21. Chavez, 2017.
22. Chavez, 2001, 2008; Santa Anna, 2002.
23. Dávila, 2008.

social and political context. At stake here was his and his partner's status as respectable, law-abiding middle-class citizens, not to be confused for temporary "immigrants" looking to take advantage of the nation.

Here Armios was very elegantly traversing the different layers of the ethno-stack that I have proposed as an approach for analysis. His initial response that he saw no separation between his hacker self and his other selves was a reflection on how his coding practices led him to strongly identify as a coder/hacker in all domains of life, at the *personal* layer of the ethno-stack. Since his reflection shows how he further understood how his identity might affect how he experienced relationships with other people, in this case his partner and the doctor, he was moving his analysis to the *interpersonal* layer of the ethno-stack. By connecting transnational constructions of race and class, issues to consider at the *sociopolitical* layer, to his subjectivities (the coder's paranoia) in the code worlds, he was thinking with me, at the *sociotechnical* level to make thinking with the computer a productive exercise. My proposal of the ethno-stack, and these specific layers of analysis for the US/México techno-Borderlands, argues that in moments like this, the code work becomes "good to think with."

[4] Rethinking Compulsive Programmers with the Ethno-Stack

Does hacking from the US/México techno-Borderlands mean that young people who navigate this uneven terrain need to be even more adept than the "average" hacker at hacking the world around them? Do they develop a desire to "game the system" more than hackers who do not share similar experiences? Thinking with the ethno-stack allowed me to think with my research participants' code work to show how subjectivities, influenced by hacker ethics and coding logics, travel across seemingly disconnected domains—the personal, interpersonal, sociopolitical, and sociotechnical layers of the ethno-stack—in subtle but meaningful ways. I ended this chapter with Armio's connections between his migrant paranoia and coder's paranoia to argue that particular subjectivities, influenced by the code worlds, are exacerbated for hackers who are not the unmarked "compulsive programmers" in the popular imagination.

While ethnographic fieldwork many times revolves around conducting participant observation, it rarely invites "participants" to actually participate in the theorizing themselves. But my chosen research group was not the typical "subaltern" group, waiting to have their voices amplified; many times they came off more as the "superalterns" other anthropologists have conducted research with. Kelty, for example, finds these superaltern subjects, in addition to being able to "speak for themselves," are vocal, loud, persistent, and loquacious.[24] Similarly, many of the young people I conducted research with over the years were happy to narrate their own practices as they figured things out; most even possessed the "problem-solving mentality" inculcated into students by university computer science and engineering programs. My invitation to reflect on how they saw particular subjectivities transferring from the code worlds to their everyday worlds was really an invitation to think with the ethno-stack, to do the theorizing with me.

My prompt received varied responses from a wide array of characters, from everyday stories about solving domestic and leisurely challenges, to amusing stories about mishaps in the dating and love worlds, to serious dilemmas where one's sociopolitical identity and livelihood are at stake. Some dreamt of code, others slept with their computers (physically or metaphorically), and others had to negotiate priorities and approaches with their loved ones—their beloved partners and their beloved code work.

Hiro's (mis)adventures in the dating worlds of San Francisco might seem humorous, or might arouse mixed reactions from readers, depending on their encounters with the treacherous terrains of love and dating, but even with his anecdotes, there was something more to him trying to game the dating system. Hiro frequently used the language of "contracts" when referring to his relationships that never seemed to crystallize. The young women who ghosted him didn't want to "enter an extended contract with him," or he had entered a 2-week contract with another woman he didn't seem to like all that much and was putting off a decision until then about whether he wanted the relationship to continue. Of course, he was borrowing this language from the labor conditions he was facing himself, where he would build prototypes for companies, some of which were only contracts of a few months, until they no longer needed his services, and some of which ran the risk of dissipating at any moment if it was for a startup company that would spontaneously report they no longer had funding. Although he was doing relatively well, financially, hustling in these short-term gigs, he found himself having to sell himself as a bundle of skills, qualities, and experiences that had to be consciously enhanced and managed in the name of Silicon Valley–inspired labor-market competitiveness.[25]

24. Kelty, 2008: 19.
25. Gershon, 2018; Martin, 1999; Urciolli, 2008.

Perhaps for Hiro, controlling this one domain of his life was a response to the ephemerality of the contract-world he was forced to navigate to gain subsistence. As other ethnographic studies have suggested, the entrepreneurial flexibility that comes to define the professional lives of many of these code workers becomes not only a mode of labor but also a way of life.[26] The entrepreneurial self enacts an entrepreneurial flexibility that "is articulated as a means by which to 'be one's own boss' means not only controlling work hours, conditions of labor, and other structural dimensions of work but also one's *feelings* about work itself, its reflection of the self, a means by which to actualize and materialize one's personhood."[27] In this sense, my research participants brought their coder selves and corresponding worker selves, the selves they constructed under the broader umbrella of hacker-entrepreneurs, to the worlds of social relationships and practices, attempting to regain some of the control they felt in those code worlds. If they could hack the coding environment, they could hack other environments. Although we saw how those ethics and logics didn't quite work out in the best way for some of my research participants, for others, it wasn't something they could really control; the coding compulsiveness or paranoia followed them wherever they went. That need for and pleasure from control that scholars have identified in compulsive programmers was not only connected to a personal relationship with the machine.

Wrestling with the ethno-stack for Mexican and Latinx code workers means taking into account how the sociopolitical is always carefully intertwined with the technical. The tinkering and problem-solving approach to life I present across this chapter is part of the broader social practice of hacking, but hacker-entrepreneurs across the US/México techno-Borderlands sometimes aim their code work precisely at the higher-order layers of the ethno-stack, as my next chapter demonstrates, when all-women organizers decide to rethink what it means to hack.

26. Freeman, 2014.
27. Freeman, 2014: 210, emphasis in the original.

[4] Abuelitas as Infrastructure

Santo Niño de Arduino amigo mío, gracias por traer a mi abuelita al hackatón de chavas. Se impresionó con nuestro invento para facilitarle la vida cuando prepara sus riquísimas salsas. Ahora ella está muy orgullosa de mí y de mis compañeras.

Aconteció en el Auditorio de la Fac. de Ingeniería, 2015

///ENG
Holy Infant of Arduino, my friend, thank you for bringing my dear grandmother to the all-girls hackathon. She was impressed with our invention to make her work easier when she prepares her delicious salsas. Now she is very proud of me and of my partners.

This happened in the School of Engineering Auditorium, 2015

[0] Firsts in the Hacker Worlds

An event advertised as the first all-women's hackathon in Latin America was held in Mexico in 2015. The event was part of a trend that emerged during my fieldwork, in which organizers rushed to create "firsts" in hackathon worlds: the first hackathon for Latinxs, the first hackathon in <insert your favorite Mexican city or favorite US metropolitan city with a high Latinx population>, the first hackathon by and for women in Mexico, and so on. The event was initially advertised as the first "for women by women" hackathon in Latin America. Because the event was partly sponsored by US universities (and later companies), and it would include presentations by female and Latina-identified professors from the United States, the event quickly became the first hackathon for Latina women, not just for Mexican women. During the boom of the hackathons in the mid-2010s, popular discourse framed inclusivity within maker/hacker groups by proposing ways to get different or "diverse" participants to join events aimed at empowering their "communities." In this chapter I explore the construction of gender at the women's hackathon to show that the management of difference within this space is intimately connected with complex ideologies of productivity, femininity, and work, as well as hackers' own ideas about how the practice of hacking can be used as a form of empowerment.

One of the organizers, Saiph, explained their impetus for the event, "A female hackathon. A Latina hackathon. We are fighting to incorporate two minoritized groups at once!"[1] Saiph was a Mexican computer science PhD student at the University of California, Santa Barbara. Like many of the organizers, she traveled freely between the US and Mexico and had experience with the hackathon circuit on both sides of the border. She also had experience with constantly shifting between "Latina" and "Mexican." Shuttling between "Latina" and "Mexican" is specific to the US/México techno-Borderlands, requiring strategic maneuvering and aligning with a longer history of the political economy of tech work that consistently shifts ethnic,

---

1. I use Saiph's real name, as requested by her, as a way to confront the invisibilization of contributions by women in her profession.

racial, and national markers to conveniently locate "cheap labor" and "passive nimble workers."[2]

On the one hand, people like Saiph were positioned by Mexican developmentalist narratives as forward-thinking, mobile, tech-savvy subjects worthy of representing Mexico across national borders. Both of her parents were professors that had raised her with what she called a "growth mindset." As she reflected on her experiences on both sides of the border, she told me, "I was privileged in coming from a family where they believed I could do anything I set my mind to and [they] did not have limiting beliefs. It is until now that I see that privilege because I have had to interact with people who have much more limiting beliefs about what Latinas can do." Saiph went on to provide anecdotes about multiple times that researchers did not believe she had implemented computing projects she had designed and worked on herself. Her father, the head of a bio-robotics laboratory at UNAM, was also one of the organizers of the hackathon, using his connections with other academic administrators to make the event a reality.

Although Saiph came from a relatively privileged upbringing and represented a select minority of the Mexican population, she was confirming her place as a "middle-class" Mexican citizen who had obtained her mobility and other freedoms by working hard. At the same time, she confirmed nationalist narratives that the software industry was the key to meritocratic progress. On the US side, meanwhile, identifying as Latina/o was for her and others like her inextricable from the shifting political economy of race. As an "emerging middle class," a homogenous Latino collective is sometimes celebrated for their coming of age in America, while the same group is at other times represented as an economic liability that takes jobs, resources, and benefits from "regular" Americans.[3] The strategic use of the Latina/o marker by people themselves points to the mobile and nomadic nature of the signifier; it functions as a political rather than a descriptive category.[4] Catherine Sue Ramírez, for example, traces the usage of "Latina" by self-proclaimed "Third World women" as a site of identification for women across the Americas to create the "geographical bridges" necessary for "borderlands feminism."[5] Thus, while resignifying the event from being a "Mexican" to a "Latina" hackathon might seem like a simple way to make the event more inclusive, participants' decisions about how to identify and relate

2. Nakamura, 2014.
3. Dávila, 2008.
4. Aparicio, 2017.
5. Ramírez, 2004a.

to one another within the hackathon were always in the context of shifting (and conflicting) markers of class, ethnicity, nation, and race constructed in relation to US–México politics.

At the same time that Saiph and others shuttled between "Mexican" and "Latina," they also found themselves strategically calibrating their alignment with the promise of empowerment that the hackathon discourse offers, as well as the push for productivity that permeated the aura of the hackathon. In this chapter I focus on how research participants at the women's hackathon position themselves in relation to these multiple demands while taking into account the gendered dimensions of the space they set out to organize. I ask how they reconfigure narratives of progress at the same time that they negotiate ideals of competitiveness and autonomy that accompany most tech entrepreneurship. How do participants figure and reconfigure structures of belonging, recognition, and visibility within sociotechnical configurations? Ultimately I argue that, in the face of results-oriented events, the women find a way to enter and transform the male-dominated worlds of hacking while they also build solidarity across generational as well as hacker and non-hacker lines, constructing subtle interventions that center and celebrate the young women's holistic hacking.

In what follows, I first highlight how participants at the women's hackathon aligned themselves with hierarchies of expertise as they addressed normative constructions of gender and femininity. By using computer development kits that allowed them to work at a higher level of the computing stack, they were able to assert their own knowledge that aligned with the "intelligent home" theme of the hackathon. Although the domestic theme of the hackathon could be framed as reproducing gendered divisions of labor, it can also become a way for women to reclaim spaces within the home as their own laboratories for experimentation, where the value of embodied practice, against dominant ways of producing knowledge, is valued.[6]

In the next section, I explore how hackers continue these negotiations of normative femininity and what it means to become empowered by the code work when they get caught up in nationalist and personal pushes for productivity. In a final section, I unpack a surprise visit by *abuelitas* (grandmothers) who taught everyone a lesson on the invisibilized labor that supports communities of hackers. By infiltrating a space normally populated by young technology-producers who represent "progress" and "modernity," their very presence re-established undervalued domestic work

6. Meneses, 2020.

as foundational for other types of work. By weaving these three threads together ethnographically, and returning to the ethno-stack, I conclude that the women's hackathon in Mexico serves as a reminder to not devalue traditionally domesticized or feminized labor, an expression of female solidarity, and an example where thinking carefully with the "bottom" layers of the ethno-stack can lead to inadvertently thinking with the higher-order, closely coupled sociopolitical layers.

[1] Stacking Empowerment, Expertise, and Coded Femininity

The representational importance of the hackathon cannot be overstated. When I began my fieldwork as the technical instructor for the six-week incubator-style course designed to teach entrepreneurship and mobile internet technology skills to Mexican students, our teaching team received hundreds of applications within days of releasing the online application to university students. Of the 32 students selected, only 4 identified as women. Similar distributions were also common in the over 20 hackathons I attended; despite the "openness" these technological communities promoted, participants at the events and spaces were overwhelmingly men. The explicit all-women dimension of this hackathon was designed to flip the script. About half of the over 300 women who registered for the event showed up and participated. Most were university students spanning a broad range of majors: engineering, computer science, math and sciences, design. There were also some younger students in attendance; the organizers had created ties with local high schools to organize transportation and get parental permission for interested girls to attend their first hackathon. They were also sisters, daughters, and granddaughters—family relationships that were always present and that became even more pronounced during the final pitches. By one of the organizer's accounts, this was the largest number of women gathered around the theme of computing technologies in contemporary Mexican history.

During a preliminary organizing meeting for the women's hackathon, veteran hackathon attendees chimed in on what features they thought would make the event successful. Some ideas aligned with what other scholars and organizers have identified as key elements of a feminist hackathon. For example, a focus on both technical and non-technical solutions to the agreed problem space, a commitment to sustaining the community of practice formed during the event after the fact, and a valuing of learning over invention in order to create a more inclusive environment composed of subject matter experts as well as marginal users.[7] The event would be held over the weekend, but designated hacking spaces would shut down overnight, allowing participants to go home and return early the next day. Parents would thus feel comfortable granting permission to their daughters to participate in an event that did not involve being out late at night. Removing the night work from the hackathon effectively intervened in cultural conventions of staying up all night that continue to be celebrated in computing communities and are many times unappealing, impractical, or unsafe for women.[8]

Additionally, in order to get as many women as possible interested in technology and software development, regardless of their background or level of expertise, organizers decided to work with Arduino kits. Arduino is an open-source electronic prototyping platform used to develop interactive electronic objects. Coders are able to program the microcontroller using higher-level code in the Arduino programming environment in their computers by downloading programs to the board via USB adapter cables. The kits prepared by the organizers for the event included components that could be used to design devices for "intelligent homes" with minimal coding experience required. In effect, hackathon participants would be working at the higher levels of the computing stack, without needing to understand lower-level stack implementations.

Ready with the Arduino kits, the organizing team decided that the overarching theme for the hackathon would be "#FixIT." This theme pointed away from the infatuation with newness that permeated hackathons and instead focused on working with existing sociotechnical objects and configurations, effectively calling attention to "hacks" that were already taking place within the home. On the one hand, then, we could clearly deconstruct the theme of "intelligent homes" as a way to reproduce gendered conventions of domesticity and divisions of labor. One could argue that designing the perfect home for women by women helps to further an ideology of women's place in the home; they're designing the perfect home so they can feel comfortable and stay there, where they belong.

However, the dichotomy between the domestic and public sphere has been critiqued as a socio-historical construction that does not represent the experiences of women in México specifically and Latin America more broadly, where domestic activities are intimately connected to and in

7. D'Ignazio et al., 2016.
8. Ensmenger, 2010.

constant interaction with markets.[9] Even with growing opportunities for women in Mexico to maintain formal part-time or full-time work outside of the home, many times they continue to use strategies already common in their communities to juggle multiple commitments related to work and family life.[10] But the hacks that occur within "subaltern design sites" such as the home never receive the recognition or resources given to those marked to be properly technologically innovative within the bounded space of hackathons and hackerspaces.[11] The name of the hackathon, #FixIT, thus served a double purpose: to ascribe value to hacking within the home as a way to "fix" an IT industry that undervalues the repetitive or domestic labor that serves as the foundation for "innovation."[12]

Thus, the women at the hackathon did not appear bogged down by the "home" aspect of the event and instead used their enthusiasm to create an intimate space and experience. Following their lead, I was often less interested in what was made in these hacker and maker spaces, seeking rather to explore how hacking functioned as a mode for these participants to intervene in and position themselves in relation to existing sociopolitical structures. In this sense I was going against one of the hacker ethics, "<3> Doing > Talking." This ethos permeated the spirit of the event, as several key moments during the event indicated that the women were not only there to learn to code, or to build intelligent homes, but to rethink modes of solidarity and expertise in the name of community empowerment.

Many of the hackathon events I participated in were aimed at empowering particular groups that had been marginalized or underrepresented in the tech industry. The hype around the events was usually framed around these groups of people becoming empowered by using technology but also by actively participating in building the technology that they might use. This was particularly important in Mexico, as government programs (and many times academic work) frequently framed technology experts as passive recipients or followers, perpetually dependent on foreign "cultures of innovation."[13] In short, the liberatory potential of the technologies, for advocates of these hackathons, rested on the premise that users could now become producers.

9. Montoya, 2002; Nieto, 2004.
10. Peterson, 2014.
11. Costanza-Chock, 2020: 140.
12. Irani, 2019: 210.
13. Medina et al., 2014.

In this view of community empowerment, marginalized populations appropriate new technologies for their own ends. In just this way, the women's hackathon, as framed by Saiph, was meant to bring together a group positioned at the intersection of different marginalized categories (Latinx, women); this newly formed community would convene, develop and appropriate new technologies, and use these to resolve issues they had judged important to their collective well-being and future livelihoods. In the case of this hackathon, those issues circulated around the setting of the home.

But learning how to code for newcomers to the hackerworlds isn't necessarily easy. It typically requires long-term education and training (whether formal or informal) and involves focused attention and practice that can take months or years—certainly more time than is provided in the space of the hackathon, or even several hackathons. A common scene that unfolds at many hackathons is one in which, after introductions and the idea/design phases of technical projects, one person—usually a man[14]—takes command of implementing a working prototype, claiming an authoritative, expert position because of his technical prowess. This often cuts against desires for the inclusive, collaborative development of expertise.

Indeed, another common scenario within hackathons is that pre-configured teams of experts arrive at the events in order to maximize their possibility of "winning" the event. In other cases, individual "experts" would rush to form the ideally structured team according to (other) expert guidelines. One "ideal" team I frequently heard described was one composed of a hacker (a person with programming skills), a hustler (a person with business skills), and a hipster (a person with marketing/design skills). In other cases, the newly formed small teams would, for the most part, effectively divide themselves up into the experts and the novices, at least along dimensions of technical proficiency. In the end, an expert programmer would be the one responsible for implementing the working project.

Christina Dunbar-Hester has found a similarly aspirational, but not quite functional, apprenticeship model present in her work with radio activists across do-it-yourself (DIY) and maker spaces.[15] Within these spaces, activists promote their vision for a self-sustaining participatory structure, one in which self-guided discovery and learning can provide a heightened sense of agency to participants, where the demystification of technology can lead to

14. Dunbar-Hester, 2014.
15. Dunbar-Hester, 2016.

a leveling of expertise through pedagogical activities. In this vision, as time progresses, novices are meant to become experts, and the field of experts within a group will increase, broadening the capability to recruit more novices, in a self-sustaining novice-expert model. Instead, Dunbar-Hester found novice participants frustrated and alienated as they attempted to learn from experts in compressed time frames.[16] She notes that although activists self-consciously tried to distance themselves from competitive and exclusionary aspects of some electronics and engineering cultures,[17] the technical pursuits they offered often ended up being intimidating and unappealing. A consistent situation across Dunbar-Hester's spaces was that "the *burden* of participation fell disproportionately on women and technical novices."[18]

I encountered similar dynamics across my research sites. The burden of *participation* often fell on novices, while the work of *implementation* was taken up by the technical experts, who often derived great satisfaction from that work, not always realizing that they might be enacting exclusionary practices. This was especially the case in hackathons that were dominated by performances of technical masculinity, as described in Chapter 2.

The participants at the all-women hackathon thus avoided these pitfalls, by, well, eliminating men, first of all, but also by aligning their expertise levels with the higher orders of the computing stack in the form of the Arduino kits. These kits allowed them to use coder tools (programs and apps) to manipulate data along with electronic components. In this way, they were able to effectively perform their burgeoning Latina or Mexican *maker* or *coder* subject positions.

The atmosphere of the 2015 women's hackathon at UNAM was thus overwhelmingly congenial and designed to foster inclusivity; the message of increasing representation in technology cultures and software development was never far from participants. The first day of the event opened with female role models who described software as "algo que te extiende, que te libera" (something that extends you, that liberates you), and attempted to convince young women that coding was far more social than popular imaginaries made it out to be. "La calidad de nuestras vidas depende de la calidad del software que construimos" (The quality of our lives depends on the quality of the software we build), another presenter confirmed. Together with other presenters, this computer scientist reminded attendees that software

---

16. Dunbar-Hester, 2016: 81.
17. See Abbate, 2012; Chun, 2013; Turkle, 1984.
18. Dunbar-Hester, 2016: 92, emphasis in the original.

development as a profession was not at odds with normative ideas of femininity—which she associated with standard roles of wife and mother, if this was how young women wanted to assume their professional and social roles. She followed a line of thinking that might be familiar from the work of such scholars as Sanjukta Mukherjee, who has argued, on the basis of her fieldwork in Bangalore, that software work has been constructed as empowering for women in parts of India precisely because its flexibility is compatible with normative feminine roles of wife and mother.[19]

In a Mexico where a growing segment of the population is claiming a middle-class identity, sometimes this means living in a safe environment surrounded by "tranquilidad."[20] Most of the time "middle-classness" means enrolling children in extracurricular activities and/or private schools, family vacations, and especially access to "modern" household appliances.[21] The speakers at the women's hackathon pushed toward this tranquilidad and comfort by promising the attendees that (1) they did not have to abandon their important family roles if they chose to become professionalized working women, and that (2) learning to code might allow them to make their home life even more comfortable and compatible with complex schedules and multitasking. Whether one could be a professional woman and a super multitasker in reality was another story. For women across the world who enter the IT workforce or related entrepreneurial initiatives, they must not only learn the skills of their trade, but more often than men, consistently convince customers and colleagues of their experience and expertise.[22]

While the women's hackathon provided the space to reconfigure ideals of gender in relation to technology, many of the circulating discourses tended to be anchored in gendered discourses of respectable femininity and putting the family first. Similarly, scholars who have conducted research with women from the Global South who enter the information technology workforce find that these women calibrate their careers and individual aspirations in relation to the well-being of their family. Some women decide to slow down their climbing "up the ladder" or give their career up altogether if they feel they are at risk of becoming bad mothers or not living up to the expectations of their in-laws.[23] They might push the boundaries

19. Mukherjee, 2008.
20. Pertierra, 2015.
21. Escobar Latapí and Pedraza Espinoza, 2010.
22. Mitter, 1995.
23. Radhakrishnan, 2008: 15.

of respectable femininity by asserting that they will not be constrained by family, but always want to be considered *of* the family.[24]

At the women's hackathon, the family, often narrowly defined within Mexico as heterosexual and patriarchal, gave burgeoning hackers a referent to negotiate normative roles of gender and femininity as they assumed "expert" roles in the software coding worlds. At stake was how they could become empowered in or by the code worlds. In the next section, I highlight how my research participants continued these negotiations in relation to nationalist pushes for productivity.

[2] Hacking Time

D'Ignazio et al. lay out premises for feminist hackathons, asserting that one important feature of such an event is that it "intentionally architects media attention in order to advocate for the issue."[25] Saiph certainly prepared for media coverage similarly. She stated in an interview, "Para mí lo importante era mostrar al mundo que tenemos mujeres creadoras; hay chicas emprendedoras y todas ellas unen sus diferentes talentos para desarrollar nuevos productos" (For me the important part is to show the world that we have women makers; entrepreneurial young women exist and all of them unite their talents to develop new products). This "productive" aspect was picked up by media outlets. An UNAM engineering professor was quoted in another news report:

> "Los proyectos que les propusimos son la base de dos cursos que ofrezco: de robots móviles y de casas inteligentes; tomamos las prácticas que hacemos ahí y las modificamos para que las trataran de hacer en dos días. Es más, me burlaba de algunos de mis alumnos y les dije: ellas lograron lo que ustedes no pudieron en un semestre."
>
> (The projects we proposed were based on two courses I offer: mobile robotics and intelligent homes; we took the classroom exercises and modified them so they could take them on in two days. Actually, I made fun of some of my students and I told them: they [women] accomplished more than you were able to in one semester.)

The professor's comments echoed much of the rhetoric that circulated during and after the event—that is, that these young women were highly

---

24. Amrute, 2016: 189.
25. D'Ignazio et al., 2016.

productive and that they were able to accomplish more than male students had in an entire semester. They were effectively mobilizing aspects of the hacker ethics "<8>always be learning" and "<9>get involved," but doing so in a way that connected with the hacker-*entrepreneurial* narratives that promoted the "promise of technology," that new technologies and corresponding entrepreneurial practices could help young people succeed quickly and dramatically.[26] And the women took this call to perform this ethos of productivity seriously, to show that they could surpass their male hacker-entrepreneur counterparts.

Whether the "all-women" dimension of the hackathon was effective or not, and whether it increased their productivity, depended on who you asked. As participants organized into teams that were composed of diverse disciplinary backgrounds—among undergraduate students there were computer engineering, mechatronics, industrial design, political science, and philosophy majors—I served as floating mentor/ethnographer, circulating between the scattered university spaces and helping with ideas, implementations, and pitches while I asked about the explicitly gendered dimensions of the event.

Mariana, a skilled and amiable computer science student whom I knew from the coding bootcamp I had administered several summers before, confirmed her excitement for the all-women structure, "En otros espacios, los hombres nos reclutan sólo para que hagamos sus apps lucir bonitas" (In other spaces, the men recruit us just to make their apps look pretty). Although the aesthetic elements of an app are of upmost importance, Mariana saw herself as a coder first and didn't appreciate being relegated to the non-coding aspects of app development. I had spent several months working with Mariana in a coding bootcamp, where many times she encountered these masculinist remarks, assumptions, and performances of technical competence. At the very least, the women's hackathon had provided a space for her to feel comfortable enough to call out the gendered power dynamics she had experienced within the unmarked (or by its un-marking actually marked as "inclusive") space where we had worked together previously. This is particularly important as women are often pressured to "speak up" in order to show their value within software design spaces.[27] The women's hackathon gave Mariana the space to speak up about the very power dynamics that framed her as an unimportant contributor in the male-dominated space.

---

26. Shankar, 2008.
27. Irani, 2019: 102.

When I asked Alicia, a recently graduated visual artist, what could have attracted more participants to the event, she responded bluntly, "más hombres" (more men). Here she was confirming that not all participants were feeling the "all-women" camaraderie the event attempted to create. T.C., a mechatronics major on the same team, disagreed with her colleague, telling me, "Las mujeres tenemos en el chip ser competitivas contra una y otra. Si fuera con hombres, decimos, 'no podemos,' y no nos activamos" (We women have it in our chip to be competitive against other women. If [the hackathon] was with men, we'd say, 'we cannot,' and we do not activate).

T.C.'s comments aligned almost perfectly with the language of the "Todos con el mismo chip" *retos* (challenges) that were part of the Mexico Conectado initiative. She claimed that her competitive "chip" was activated within the space of the all-women hackathon. While the government initiative saw the chip as a way to instill an "innovative culture" in its young citizens that would help Mexico scale, "develop," "modernize", and appear economically competitive as it played catch up with the rest of the world, T.C. saw her activated chip as a means for women to "catch up" to their male counterparts, or perhaps surpass them, as the professor commented. This ethos of productivity, of looking busy, and being explicit about how one was spending their time, permeated the hackathon in unique ways. As anthropologists have shown, the question of how young women from the Global South spend their time, and whether it's leisure, pleasure, or "work," is a critical site to investigate constructions of gender, productivity, and social change.[28]

The young women thus took the call to be productive within the hackathon—whether this came from the activated chip or not—very seriously. Many I spoke to agreed that a second panel of speakers that was planned during the opening of the event should be cancelled; they claimed that these extra presentations were effectively "wasting their time." Those who had been to other hackathons felt that since these other events did not have any additional presentations, it was unfair that these extra time commitments cut into their hacking time. Time constraints created anxieties and re-shuffled priorities among the groups. I was interested in learning more from a group of three hackers who had been diligently creating a cardboard mockup of an entire home with what seemed to be flashing lights in different sections of the home. As I approached the team to develop rapport in order to ask some open-ended questions, one team member told me to go

28. Amrute, 2016; Fleming, 2018; Krishnan, 2018.

away—I was taking up their time. I didn't take offense to being referred to as a "waste of time." I was conscious that I might be framed as an intruder, particularly by young women who were new to a hackathon space and had not worked with me in other hackathons or the MIT bootcamp at UNAM.

I was able to speak more candidly (or at least at all!) with Ío, though, a mechatronics student and graphic design enthusiast with whom I had built confidence over the years. Her team was working on a prototype of a baby robot that would crawl alongside infants and keep them entertained while the human mother worked on tasks other than childcare. She admitted that most of her team knew each other from their mobile robots course at UNAM and that this project seemed like something they could implement in the short time provided using skills they already had. "Mi compañera del grupo tiene una hermanita bebé y dice que se entretiene gateando hacía una Chihuahuita que tienen. Al robot lo podemos controlar; al chihuahua no" (My team member has a baby sister and she says that she entertains herself by crawling toward a Chihuahua dog they have. The robot we can control; the Chihuahua we can't). The team understood the safety features and concerns that such a robot would entail, but after all, these were working prototypes.

I also told Ío about the group who told me to go away and wanted her impressions about my presence at the hackathon. She laughed about the group's directness, and about me being at the hackathon, she said, "En tu caso tal vez está bien porque estudias hackathones. Pero ser antropólogo no te hace diferente tampoco. Al final eres un hombre que va por ahí, criticando, siendo el especialista . . . No puedes entender lo que es estar aquí celebrando con mujeres, celebrando estar juntas" (In your case I guess it's okay since you study hackathons. But being an anthropologist doesn't make you different either. In the end you're just a man going around, criticizing, being the specialist . . . You can't understand what it means to be here celebrating women, celebrating being together).

Taking my cue to be a "productive" member of the hackathon instead of asking and trying to understand so much, I worked with teams to put together their final pitches for their prototype designs. We perfected speeches and added the final bells and whistles to the presentations. They would only have 3 minutes and 30 seconds each to present, so getting the presentation done quickly and efficiently was as important as how it looked. Overlooking us in the engineering building was a commonplace "Silicon Valley map." The dated poster showed an illustrated map of the San Jose, California, region with dozens of company names superimposed on the map, representing the company's geographic location. The poster was a sobering reminder that this hackathon was meant to mimic the high-paced, competitive cycles of technology-driven capitalism that characterized Silicon Valley.[29]

It was also a reminder that productivity as a unit of measure, rather than a calling, comes from technologies of efficiency that came to prominence with an industrial-managerial orientation that established the language and practice of productivity. The capitalist hijacking of productivity might be implicit in the poster, but one thing that might be difficult to discern from the visual is that the history of productivity is gendered. Melissa Gregg shows how the gradual retreat of the office secretary in the service of others' productivity has been offset by technologies that continue the same tradition of entitled delegation.[30] For an elite group of professionals, new technologies provide the sociotechnical infrastructures that enable them to *practice* productivity. Whether new technologies would make life easier for the women at the hackathon, and how they might position themselves in relation to the home vs. professional life dilemma, was up for debate.

Also up for negotiation was whether the all-women design of the hackathon was a positive or negative aspect of the event, and whether some presentations, or even my presence, was "wasting their time." But one thing that all participants did agree on, and made evident, was that they were excited and nervous for their final presentations—the spotlight would now be on them to show what they had spent the weekend *making*.

[3] Re-coding Infrastructures of Work and Progress

The nervousness that the young women demonstrated during their final pitches was warranted, for different reasons. For one, the event was not entirely all women. As different teams took the stage to present their home improvements, the young men who had participated as mentors during the weekend stood alongside the walls of the auditorium, arms crossed, dressed with black Google shirts to distinguish themselves and claim authority to the probing gazes they directed toward the stage. In one particularly condescending comment, one of the mentors voiced a demeaning, "Awwwwwww," as Ío's group presented their crawling baby prototype. One of her teammates smiled uncomfortably. Some young women in the audience rolled their eyes.

---

29. See Jones et al., 2015.
30. Gregg, 2018: 5.

The presence and help from the young men resembled similar dynamics to those Leslie Salzinger finds in the Juárez maquiladoras of the early 1990s.[31] Salzinger argues that as young women from the Global South became neoliberalism's preferred workforce, the industrial shop floor became a space for contesting the feminization of work.[32] Male workers attempted to reclaim the inherent masculinity of the work by constantly teasing, crowding, and "helping" their female co-workers in an effort to put their inadequacy on display. While the hackathons of the 2010s might appear to be a world removed from the industrial maquiladoras of the 1990s, the underlying frameworks that constantly "test" the skills of and ultimately marginalize women workers stand the test of time, as the "well-intentioned" mentor-helpers reminded us.

Under the gaze of these male helpers, then, the teams continued to pitch their projects to an attentive audience. The first-place team, "On-the-Go Mom," used its Arduino kit to present prototypes for intelligent sensors meant to detect when food supplies were running out. They wowed the audience with their pitch, which pointed to the many features and benefits of such a system: a user could view the stock of their kitchen in real time from the supermarket, it could help save families money by not buying unnecessary items, and a recommender protocol would even suggest food recipes using food items in stock. Mariana's team, "Easy PetCare," placed second with a system that would track their pets around a set perimeter, alerting the user when the pet had left the specified boundaries. And the third-place team, "Keypeer," developed a key holding device that would alert users when all members had safely arrived home. Other salient themes across the prototype presentations were home security, pet care, disabled care, and automation of everyday household tasks. As the teams progressively gained confidence, their nervousness eased, in no small part because of the presence of a new group of audience members who silently but elegantly started to arrive on scene: mothers and grandmothers who had come to witness the presentations and cheer on their daughters and granddaughters.

The surprise visit from *mamás* and *abuelitas* certainly added a "different" element to the hackathon. I had never seen family members come to cheer on their burgeoning hacker-entrepreneurs at hackathons in the US, for example. Neither had I seen them at the male-dominated hackathons in

31. Salzinger, 2016. "Maquilas" or "maquiladoras" are assembly plants that have been established in Mexico since 1965 as part of the Border Industrialization Program, employing thousands of young Mexican women to assemble, manufacture, or repair goods for export to US or other foreign consumer markets (see Iglesias-Prieto, 2017).

32. Salzinger, 2016.

Mexico. The family ambiance quickly turned the hackathon into an (even more) hospitable event, with lively cheering for family members and extra commentary from the sidelines. As I circled around the room to speak with the family members, I asked one grandmother what she thought about her granddaughter's intelligent home prototype. "Lo que no hizo en todo el semestre lo hizo en dos días." (What she didn't do all semester she did in two days.)

While the abuelita's comment was in line with the comments regarding productivity that had circulated throughout the event, the very presence of these older women signaled a departure from the "grandmother" discourse that normally circulates in computing worlds. That is, in addition to the performances of technical masculinity that many times intimidate women from participating in technical cultures, it's common among computer experts to use gendered language specifically to make women feel like outsiders. In some FLOSS (Free/Libre and Open Source Software) communities, for example, the term "Aunt Tillie" is deployed to refer to someone who is not tech-savvy.[33] Across communities of computing experts, it's common to hear subjects urge you to build a system or interface that is "easy enough for your mom/girlfriend/grandmother" to use or figure out.[34] The complementary guidelines for a clear explanation of a complex system is to "describe it as if you're explaining it to your grandmother." Ironically, in this imaginary, it is always the grandmother (never the grandfather) who is the willing recipient of (most likely) boring technical explanations. One can imagine the grandmother waiting at home, ready to listen and provide encouraging words on your work in progress. Perhaps in these imaginaries the grandfathers, uncles, and brothers are busy in the co-working spaces becoming the "modern," tech-savvy, mobile workers of tomorrow. Of course, the supposedly unmarked hackathon space that takes center stage in the narrative of the ascending Aztec Tiger economy overwhelmingly features men in its protagonist roles.

The all-women's hackathon space, and perhaps even the "intelligent home" theme of the event, gave the abuelitas the opportunity and confidence to assume roles as active audience members. After the event, I asked Ío what she thought about her abuelita coming to watch her pitch. "Pues me da

33. See, for example, the writings of Eric Raymond, a prominent advocate of open-source software: https://web.archive.org/web/20210418034618/http://www.linuxtoday.com/developer/2002011401620OPKN (accessed February 9, 2023).

34. Thanks much to Luis Felipe Murillo and Sareeta Amrute for bringing to my attention these examples.

emoción. Ella es la que ayuda en todo del día a día. Ella es la que se encarga de todo" (Well, I get excited. She's the one that helps with everything in the day to day. She is the one that is in charge of everything.) With recognitions like this by the hackathon participants, the abuelitas became more than cheerful audience members. In a space usually reserved for young makers who understand "new" technologies, they claimed their space within "progress" and re-asserted undervalued domestic work as foundational for other types of work.[35] Instead of the grandmothers becoming the referents for non-technical persons, novices who don't understand and thus need to have complex infrastructures explained in very basic terms, here the abuelitas asserted their place as the very backbone of these sociotechnical configurations, the necessary defaults for the making and corresponding discourse of technological progress to occur in the first place.

The very presence of the abuelitas, mothers, and families on the sidelines effectively displaced the technical (male) mentors, reminding everyone of the "other" work required to make the projects being presented. The first tactic for a postcolonial computing reads, "When we see a technoscientific object, we investigate its contingency not only locally but in the infrastructures, assemblages, and political economies that are the conditions of its possibility."[36] If the projects presented at the women's hackathon are the technoscientific projects to unpack, in this case, we would have to explore the hierarchies of labor involved in constructing the Arduino chips. The university and its laboratories where the hackathon took place could also be analyzed as infrastructures that required manual labor to construct and maintain, which could make us dig deeper into the classed and racialized labor economy in Mexico, connecting these processes to overarching political economic assemblages.

The presence of the abuelitas, however, shifts attention to local and immediate. If the young men in Google shirts observing from the sidelines represent the technical infrastructures these young women are purportedly appropriating and the global economy that they are meant to join, then the abuelitas intervene to remind us of the work required so that these young women could participate in the event in the first place. At the same time, the abuelitas remind us that while these young men were present being the "expert" moderators of the event, they were many times absent in other

35. Gil (2022) further explores how a fab lab in São Paulo makes the space inclusive for the co-presence of cross-generational makers/hackers, as women bring their children to the space.
    36. Philip et al., 2012: 10.

domains of family life, in terms of encouragement and support; the cheering family members on the sidelines were overwhelmingly women (grandmothers and mothers) and children. While some of the abuelitas encouraged their granddaughters to take up the productive discourse of the event, their very presence reminded them not to diminish "other" labor in the name of becoming the productive workers of tomorrow.

After the winners were announced and prizes were distributed, celebrations ensued as the hackathon participants joined their families, some of whom had brought flowers and homemade food to reward their weekend efforts at the hackathon. I approached Mariana as she celebrated with her team, which won second place for "Easy PetCare," to congratulate her. "¿Qué te parecerion las presentaciones finales?" (How did you feel about the final presentations?) I asked her. "Pues, están más bonitas, ¿no?" (Well, they're prettier, aren't they?) Her response was clearly ironic, in relation to the earlier remark she had made about men at other events recruiting her to make their apps "look pretty."

Mariana's response also functioned as a way to remind me that perhaps my "end results" type question was in line with a masculinist, results-oriented technical outlook of the world that was missing the importance of what was really going on at the event: solidarity and interventions along gender lines, as peers, mentors, and family celebrated the young women's weekend efforts. I was part of this group, no doubt, but also an outsider in more ways than one. Here, at the peripheries of the event, in my attempted ethnographic encounters/interventions, I myself was an "exception" (to evoke a coding concept) to what was going on and what was important within the confines of this ritual, a hackathon that had been appropriated for means and negotiations much different than perhaps its original Silicon Valley, US-based variants had been created for.

A straightforward read of the "intelligent home" design competition at the women's hackathon in Mexico might appear contradictory. How can any of these technologies be liberating, as some of the speakers preached, if they assume and presuppose that women must maintain their interventions within the domestic space and their "empowerment" within the confines of the hackathon? This interpretation presumes that it isn't something the women (organizers and participants) had already discussed, and that it isn't something that was debated within the space of the hackathon itself. It would also be homogenizing, as the varied responses by my respondents on the "all-women" component of the event demonstrate, they all had different takes on the matter. Finally, as Mariana's ironic comment and posturing shows, there might be spaces of critique within the hackathon that this male ethnographer did not have access to—rightly so.

[4] Tight Coupling of the Ethno-Stack

During one of our preliminary organizing sessions for the women's hackathon, one of the male veteran hackers connected to our video conference call from a treadmill. He spent the hour-long session chiming in with comments while he sprinted in his gym clothes. During our introductions, organizers took turns establishing their experience, not only by invoking the software languages and platforms with which they were familiar, but by enumerating the hackathon events they had previously attended. This spirit of productivity, of making the most with one's time while juggling multiple demands, clearly infiltrated the space of the hackathon. Even if there was a deliberate attempt to depart from the "boy's culture" usually associated with technical expertise, or the technical masculinity that precedes computing, the media coverage, as well as the burgeoning Latina hackers, took seriously the call to avoid "wasting" time.

The image of the young man running on his treadmill during the meeting also coincides with the overarching narrative of "catching up" that tends to permeate events designed to empower communities through technology. In a history of computing that forgets "programming" was initially a woman's job and a computing industry that has treated women of color as ideal workers to later invisibilize their contributions and labor,[37] the stakes for Mexican and Latinx women coders are particularly high. In their case, "catching up" means negotiating nationalist narratives of productivity as Global South countries attempt to perform their departure from "developing country" status, as well as reconfiguring normative feminine roles instituted by nationalist constructions of the middle-class family. Shuttling between "Latina" and "Mexican" requires strategic maneuvering that points to how the management of difference is intimately connected with complex ideologies of productivity and work as well as nationalist and developmentalist projects.

Whether they identified as Mexican or Latina, then, at the women's hackathon, some hackers praised the liberatory potential of new technologies; other women enjoyed the competitiveness and productive time the hackathon fostered; other women resisted the explicit othering and called

37. Light, 1999; Nakamura, 2014; Ramírez, 2004b.

for more male hackers; other women (the abuelitas) made their presence felt and performed their solidarity with the makers-in-the-making, regardless of what was being made at this hackathon. Most importantly, from a space usually reserved for young makers to learn new, "modern" technologies, the abuelitas asserted their place as the very backbone of these sociotechnical configurations, the necessary defaults for the making and corresponding discourse of technological progress to occur in the first place. As proponents of the Pluriverse, a world in favor of a multiplicity of possible worlds, remind us:

> In both the global North and South, it is most often ordinary caregiving mothers and grandmothers who join this entanglement—defending and reconstituting communal ways of being and place-based forms of autonomy. In doing so, they . . . draw on non-patriarchal ways of doing, being, and knowing. They invite participation, collaboration, respect and mutual acceptance, and horizontality.[38]

To allow for some of these intergenerational and inter-hacker interventions to happen, the organizers of the event strategically used Arduino kits to place the young women at the higher level of the computing stack, from where they could assert their own expertise without having to learn how to navigate the lower layers of computing abstraction. This allowed the women time to concentrate on "hacking the home," a theme which, at first take, can be construed as re-instating gendered divisions of what one is able to hack, but can ultimately represent a joyous experience for women—feminized practices such as domestic duties are usually central activities in feminist hackerspaces because women feel pride in hacking devices they are familiar with.[39] Moreover, this also allowed women to think with layers of the ethno-stack, mainly the personal and the interpersonal layers. They made space for the abuelitas to make an appearance at the hackathon, and some participants even thought about their mothers and abuelitas as they designed their prototypes. To think carefully within these layers, to think with *care*, about their families and issues important to them effectively opened up space for the intervention to point to the "higher-order" layers of the ethno-stack, the sociopolitical layer that might be concerned with the people and communities left out across the transnational political economy of programming work. Far from the loose coupling approach to writing code, where elements

---

38. Kothari et al., 2019: xxxiv.
39. Dunbar-Hester, 2020: 108.

or entire layers should not be concerned with each others' implementations, the *thinking with* here showed that there is no separation, that the layers of ethno-stack might be, or should be, tightly coupled. The personal, interpersonal, and sociopolitical, in this case, all go hand in hand. The women at the hackathon demonstrated that perhaps the sociopolitical is better addressed at the personal or interpersonal layers. My next chapter continues with this approach, thinking with the tightly coupled layers of the ethno-stack, and returning to think closely with the sociotechnical layer, to explore how iteration becomes central to the construction of a people.

[5] Making Latinx Makers

Doy infinitas gracias a la Virgencita del Agile programming, pues después de suficientes loops, iteraciones, y renovaciones, podré ser un Latinx Maker hecho y derecho.

Guada-Loopero, San Francisco 2017

///ENG
I give infinite thanks to the sweet Virgin of Agile programming, for after enough loops, iterations, and renewals, I'll be able to become a respectable Latinx Maker.

Guada-Loopero, San Francisco 2017

[O] Life as Software

Ten years from now, who knows, maybe management is completely overhauled, and it's a different group of people, and they want to sell all of our data to data brokers—nothing is stopping that. This is why policies are important, and I've been working hard to protect ourselves from future versions of ourselves.

—JESSEE, CODER AT AN OPEN-SOURCE SOFTWARE COMPANY

At the time of this interview, Jessee was working for a startup that created a popular open-source coding platform. The startup subsequently became a very profitable company and its technical infrastructure was highly respected by hackers everywhere, since it provided the social platform where coders could collaborate on software projects and gain recognition for their personal contributions to new and existing programs. Most felt that this was what a successful company run by hackers should look like; the company's "success," in that it was generating millions of dollars in revenue, had purportedly not compromised the hacker ethos of the founders or its employees.

Eventually Jessee launched his own tech startup, and I asked him about what his daily life as a Latinx CEO in Silicon Valley looked like. He gave a pretty standard response for an active leader of a company: start with exercise, cut out a segment of the day for "proactive" time (advance new projects you're working on), cut out a shorter segment for reactive time (respond to important emails), spend some time in a Slack channel with your lieutenants, and end the day by reflecting and making a list of things you plan to accomplish the next day.[1] But his weekends were a lot more interesting. This was when he connected with a productivity advisor/guru to reflect on his week and think creatively about how his next week might be more productive, more efficient—a better version of the previous week. Just as he had been "working hard to protect ourselves from future versions of ourselves," Jessee's general orientation toward life was one of perpetual iteration and renewal. His weekends were spent *thinking through* with *others*—he demonstrated his penchant for iterative thinking and discursive prototyping. His

1. Slack is a popular messaging app created to help distributed teams communicate efficiently.

company vision proudly announced that it was "for the coder by the coder," and his embodied practices correspondingly mimicked the sprints, iterations, and openness of software methodologies. Each iteration was meant to be better than the previous; each new version was open to contributions and suggestions from other members. Better software. Better prototypes. Better people. Better futures.

Within the space of the hackathon, the prototype becomes a representation of this iterative way of being in the world. The prototype functions as a preview of things to come, a demonstration of what has been worked on but is always a work-in-progress, always open to suggestions and ready for renewals. In this chapter I explore how *making* prototypes becomes entangled with the *making* of particular groups. I show the importance of performance and participation as components of making by focusing on the subtle constructions of global Latinidad through a series of hackathon events named Migrahack that took place in both the US and Mexico.[2] "Migra" refers to US immigration authorities. I demonstrate that *making* and *participating* become ways of "thinking through", imagining, and negotiating, in community, what Latinidad looks like.

Jessee's iterative way of life is one valued and many times mimicked by other US/México hacker-entrepreneurs. On the Mexico side, hackers put in code work while they're constructed as always "in-the-making" and always in the process of "becoming." They are purportedly stuck between the "specter of the Indian",[3] the primitive and the uncivilized that forever deters them from becoming full-fledged "modern," respectable "mestizo" citizens.[4] On the US side, the hackathon becomes another space where they are supposed to resolve their positions between assimilation and multiculturalism while navigating the shifting politics of racialization. As anthropologists and sociologists who take Latinidad as an object of study have shown, Latinxs are encouraged by institutions to promote their racial identity and "be themselves," but only within the confines of respectable US citizenship, which many times equate to white middle-class values.[5] Latinxs must negotiate

2. Migrahack hackathons have taken place in Los Angeles, Chicago, Mexico City, Tucson, Denver, and Toronto. While they're focused on US/Mexico border issues, the events take place anywhere where people want to attempt to resolve related issues. Not surprisingly, the events drew large numbers of Latinx-identified participants.

3. Leal Martínez, 2016.

4. Lomnitz, 2001; Yeh, 2015.

5. Pérez, 2015; Rios, 2011; Rosa, 2019. More broadly, these studies are part of scholarship that has looked at what it means for Latinx youth to become "respectable citizens" in relation to neoliberal subject formation and the politics of race (Cacho, 2012; García, 2012; Ramos-Zayas, 2012).

which type of citizens they want to become as popular discourses frame them as either a "threat" to the nation, deploying metaphors of contagion, invasion, and disease,[6] or as properly ordered Latinx subjects grounded in distinctive American values of hard work and family.[7] Outside of the hackathon, then, Latinidad seems to be already constructed in ways that preclude a more liberatory form of transnational Latinidad. What about inside of the hackathon, where the point is to construct new versions of objects, to continuously prototype new programs and new selves?

At the Migrahack, "Latinxs," or Mexicans and people who identify as Latinx (sometimes Mexicans themselves) on both sides of the border, are summoned to learn to code and use their bourgeoning technical skills to prototype ways to empower their community. Community here already assumes that the Migrahackers belong to a multiplicity of communities. On the one hand, hackathon participants frequently mention the creation of a "global community" of hackers. If you participate in a hackathon, you're automatically admitted into the family of hackers across the globe—you are now part of the family; I heard several hackathon organizers begin their introductions to the events in this way. This call parallels the mission of popular social networking sites such as Facebook, whose proponents use "community" to mean collective identity, and who promise us that this community will lead us back to the original "community," one based on physical proximity and shared institutions.[8] Attempts to define or describe any "community" from a research perspective resemble classic ethnographies by anthropologists' who bounded particular communities in order to present them as their "field" or ethnographic unit of analysis.[9] Here, however, the creation of the "community" becomes important precisely because of the shifting politics of racialization across borders in which my research participants find themselves enmeshed. The term "community" and the term "Latinx" are many times strategically used by US/México hackers themselves, to call on subjects who will occupy the new iterations of Latinidad, those who will learn to code and use their code work to resolve political problems.

I thus highlight how communities from both sides of the US/México border put in the code work to resolve issues that they deem important to their livelihoods at the same time that they put in the cultural work necessary to perform and construct their Latinidad. What's at stake for newcomers

6. Chavez, 2001, 2008; Santa Anna, 2002.
7. Dávila, 2008.
8. Boellstorff, 2017.
9. For critiques of these approaches, see Gupta and Ferguson, 1997.

to this hackathon is not only becoming a maker, but also a *Latinx* maker. Popular discourse thinks about racial diversity within maker/hacker groups by proposing ways to get different or "diverse" participants to join events aimed at empowering their communities; here I explore how members of racialized groups are called upon to construct and manage these differences themselves within hackerspaces and "maker" formations. To explore these transnational dynamics, I bring together scholarship on prototypes and participatory models with work from Latinx Studies on constructions and mobilizations of Latinidad. I argue that making prototypes becomes a way of making hypothetical versions of global Latinidad, helping participants to think through—as members of a hypothetically never finished community—issues related to US–México relations, border "security," and (im)migration.

[1] Prototypes

The prototype is an essential component for creating futures by using software at the hackathon. The expectation is not that one will complete a polished working product, but that teams will present to a panel of judges their works-in-progress, a preview of things to come. This methodology stems directly from software development processes, where developers release "beta" versions of their programs, receive feedback from users and other developers, and use this information to iterate on their designs and implementations, in order to get closer to a final design, to approach a product/project that might overstep its "prototype" stage and be ready for a public. The stage at which the prototype is released can range from very early (e.g., asking for input from users when user-interface designs are drawn up on cardboard mock-ups) to very close to "completion" (e.g., adding or deleting bells and whistles on projects that are almost ready to be launched). Indeed, a hallmark of open-source code is that it is technically always in beta—releasing any version of your code is an invitation for others to contribute code that might add their own features to your program or even to contribute code that might re-implement a feature of your program with more robust or more elegant code.

Fred Turner traces the emergence of the prototype in professional software development to a 1990s manual, *Prototyping*, that redefined the initial "requirements" phase of system development. "The prototype could become an object, like an architect's model, around which engineers and clients could gather and through which they could articulate their needs to one another. It would speed development, improve communication, and help all parties arrive at a better definition of requirements for the system."[10] Alberto Corsín Jiménez connects the rise of prototyping as a form of cultural discourse that emerged in design and engineering circles with experimental moments in more academic domains, specifically those that engaged the social studies of science and critical theory. He conceptualizes the prototype as both material culture and sociological theory:

> Prototyping as something that happens *to* social relationships when one approaches the craft and agency of objects in particular ways. A cultural moment, then, when the prototype stands for the mutual prefiguration of objects and sociality; when objects and social relationships are recursively parenthesised, now as protos, now as types, with respect to each other.[11]

In this way, prototypes are situated within the larger field of prefiguration as "things-that-are-not-quite-objects-yet." If information technologies are the "socio-material apparatuses that align themselves into more or less coherent and durable forms,"[12] then the prototype is the manifestation of these apparatuses, or configurations. The *social* aspects of prototyping become intertwined with, and as important as, their hypothetical nature.

Viewed this way, the study of new technologies shifts from thinking of "inventions," understood as singular events, to interest in ongoing social practices of assembly, demonstration, and performance. In the case of the Migrahack, each completed (or not completed) prototype is *assembled* with an aim to resolve pressing societal issues and injustices, in this case those that crystallize around a focus to hack la migra. And we know that as the prototype comes into being, it is *demonstrated* to a group of experts that will subsequently determine which social-technical manifestation is more effective at delivering a message that might prompt others to act, to spring into collective action for a just cause. But how is all of this being *performed* at the Migrahack?

[2] Code Work == Migra Work

Similar to the spirit and ethos that filled the Hack CDMX and other hackathons across the US/México border, participants at the Migrahack fill the space with contagious excitement. They are here to discover the potential

---

10. Turner, 2016: 258.
11. Corsín Jiménez, 2014: 383, emphasis in the original.
12. Suchman et al., 2002: 163.

of new technologies, foster burgeoning collaborations, and resolve pressing societal problems. Like other hackathons, the event's aim is to get diverse folks with different abilities together for a weekend and have them use technological tools—in this case, preferably of the open-source variety—to create projects, or prototypes for projects, that visualize data, tell stories, and propel citizens into action. For many in attendance at the Migrahack, it is their first time attending a hackathon. As publicity for the event succinctly states, "Most journalists and community members have never been involved in a hackathon. Most programmers have never been involved in immigration issues. Migrahack brings them together." Monica, an executive professional of a media company that publishes newspapers and websites in cities with a large Latino population, encapsulates much of the vision and enthusiasm for the hackathon with her comments:

> Hackathons are remarkable in that they bring the power of technology, programming, engineers, to think about ways to solve social problems, and combines it with journalism, and journalism that focuses on the immigrant community, and if we can pull that data and not just tell stories from it but then provide people solutions and to force accountability—I think that's what so powerful that comes out of this.

It's easy to find publicity about the Migrahack. There were many documenters in attendance at the various instances of the event. In fact, documenting the *performance* and *participation* becomes crucial for this hackathon.

This smaller-scale, more intimate hackathon experience distinguishes itself quite explicitly from other hackathon events in that the participants are there not only in the name of creating innovative solutions to abstract societal problems, but also to address very specific problems and politics they are familiar with and that affect them personally. Unlike hypothetical problems at other hackathons, the problems here are real. The personal, interpersonal, and sociopolitical layers of the ethno-stack instance here are tightly coupled.

This tight coupling becomes apparent as teams begin their prototyping at the Denver instance of the Migrahack. Of the twenty or so teams that begin to form, at least half of them focus on dispelling immigration myths and attempting to tell migrant stories. Even before considering what technologies, media formats, or datasets will be used, the prototypes that emerge show that participants are intimately familiar with these stories. One team prototypes their project "An Immigrant's Journey" by hand sketching a multi-page chaotic flow chart that details the treacherous paths and impossible decisions migrants need to make to arrive in the US. From gang persecution to robberies to detention centers to ludicrous court processes to perilous existences as migrants, the team clearly had firsthand experiences. For those who didn't, I heard several participants calling or connecting with people they personally knew, to contribute their authentic stories. These ended up across other teams, who visualized poignant stories about unaccompanied minors, undocumented and unlicensed drivers, or the cumbersome paths to gaining refugee or asylum-seeker status. Some team names told the story—"Desesperación" (Desperation).

Far from the loose coupling that modular code design calls for, the ethno-stack being instantiated at the Migrahack thus calls for a particularly intimate and passionate coupling. This hackathon purports to not be of the standard "make the world a better place" variety that is easy to criticize and categorize as idealistic and/or naive. Participants at the Migrahack arrived with a mission to empower communities they felt a close connection to, and in the process empower themselves. Both the Migrahack publicity and participants such as Monica freely circulate the term "community." The Migrahack organizers are keen to call attention to the "community" members who can become part of the problem-solvers using code, and the participants readily reify the "community" they will be helping. The lure of the hackathon is that these communities might even overlap, creating an even more powerful collective.

Cindy, who worked for an immigrant and refugee rights organization, is eager to respond to the call. "A lot of the work I've done has focused on advocacy around immigrant rights issues, so I was hoping that coming here I would meet other people who are interested in similar issues as I am but also in creating a solution to the problems that I've seen," she says. Just as folks who have been working on immigrants' rights issues feel naturally drawn to the space, so do the hackers who have always felt close to technology. Armios tells me, "I've always been interested in immigration issues, and it's always been something that's very close to me, and naturally I've always been into technology. To a certain extent I'm doing what I feel I'm supposed to do. It's something that just—I feel I'm driven to do it." Life experiences bring diverse people to the hackathon, and these same experiences bring the layers of the ethno-stack into closer alignment.

From the outset, this hackathon seemed to have a more *participatory* feel to it than the other events where I had spent time. It seemed like people wore their hacker ethics proudly, especially "<9> Get involved" and "<5> Solve problems." One might say that there was an unspoken commitment to make sure that everyone who wanted to participate was actually able to do so. This was evident in the very structure of the event; unlike many other hackathons, a full day was devoted to workshops for "newbies" to learn new technical skills and vocabulary (data mining, mapping, Fusion Tables) in order to be able to scrape the web and access relevant immigration data, while more experienced users updated their technical repertoire with workshops on new (open-source) software to manipulate datasets and visualize newly acquired and cleansed data.

After the workshops, willing participants did quick pitches of about 1 minute of their proposed projects to speed up the recruitment process; floating participants quickly joined teams and got down to prototyping. All of this happened very quickly—they had no choice. Like at other hackathons, the clock ticked away until the final celebratory moment came to present a working demo to a panel of judges. Although the advocacy projects at Migrahack looked very different than projects at other hackathons, especially from the techno-entrepreneurial startup worlds, the final pitch session is what really bound them together. In approximately 48 hours, teams had to use their newly acquired datasets and tools to develop a technological prototype (of an app, platform, video, visualization, or other creative media genre) aimed toward raising awareness or helping solve an issue related to the immigrant population. Veteran hackers joined newly initiated hackers. US-based hackers made it to the first instance of the hackathon in Mexico City. Mexico-based hackers who had the privilege to travel to the US were sometimes found in the US versions of the event, in Los Angeles, Denver, or Chicago.

Across the US/México techno-borderlands, they worked together to create projects that used open data to tell stories about border militarization; immigrant detentions and deportations; migrant access to healthcare (on both sides of the border); and retained belongings at the border.

For this latter project, developed at the event held in Mexico, the team was composed of members of a nonprofit which documented human rights violations within US immigration centers. Their aim was to bring attention to a violation that occurred frequently at these detention centers but that was rarely discussed: the retention of whatever meager belongings have survived migrant journeys. Migrants were frequently unaware that they could ask for the return of their belongings, so they often get deported without them. "They cannot get jobs; they risk being arrested for not carrying official identification. If their relatives send them money, they cannot make the withdrawal in the bank, and if you ask the favor to someone else, they run the risk of the money being stolen," team member Cindy says. This project, which used an animated video to convey this data to the public, received honorable mention at the Mexico Migrahack event.

On the other side of the border, in a place far removed from the "border" but where people are fully aware of border issues,[13] the winning "Finding Care" project at the Chicago Migrahack used data from the Affordable Care Act to visualize unequal access to healthcare. To make their pitch even more compelling, they combined this data with the story of a 24-year old undocumented Chicago migrant who needed a kidney transplant but was ineligible to be placed on the organ transplant list because of his undocumented status. Their web-based presentation has audiences listen to Jorge's first-person account via a short audio clip while they view a large photograph of a young man, hand over his unkempt hair in anguish, looking down to the ground with despair—a completely desesperado Jorge. Over his image, the succinct quote that tells the whole story: "I'm being denied life."

This project, along with many of the videos, animation, and data visualizations that were produced at the events, corresponded to an emerging form of participatory advocacy media that was not just about an "issue" but also about a particular campaign aimed at resolving the issue. In this sense, the construction of the "issue" mimics the way experts and "reformers" specify problems that need to be fixed or improved. These social justice interventions are first conceived through the process of problematization, where the issue is first outlined and specified as something that needs fixing,[14] and then "rendered technical,"[15] the process by which experts conceptualize the worlds as ripe for "intervention" with the technological instruments they have at hand or are in the process of designing.[16] Again, in this case, these issues were more closely coupled across these technical layers.

With the issue to be resolved carefully articulated, many of the apps and projects at the Migrahack thus resembled this genre of advocacy media that was explicitly non-neutral, and refused to provide a closed narrative or structure, with the intention to invite audience members to "meet the victims," to become aware of the (many times) gruesome facts, and most importantly, to *act*.[17] These media used techniques of "audience engagement" to tell concerned citizens how to get involved, who to connect with,

13. For discussion of "Mexican Chicago," see De Genova, 2005.
14. Li, 2007.
15. Mitchell, 2002; Rose, 1999.
16. Sims, 2017.
17. Gregory, 2012: 526.

and where to sign up.[18] The panel of judges at this Migrahack event clearly had an eye for this form of media advocacy, as they commended the Finding Care project with the following text:

> Coherent, elegant narrative with lots of points of departure. Triggers questions for further research. Polished production in short time frame with simple, effective data visualization. Would love to see calls to action—links to advocacy groups, reporting on pending legislation, and so on.

Despite the call for more audience engagement, more explicit "calls to action," the judges commented nevertheless on the effectiveness and elegance of the visualization and, more importantly, on the ability of the team to develop the project in a short time frame.

But the real idea behind any hackathon, including the Migrahack, is to produce a working prototype, or an MVP (minimal viable product),[19] in a limited time, under constraints that mimic Silicon Valley–style free-market cycles.[20] The time pressure was not lost on the hackathon participants, and it was reflected in their overarching feelings: in their desires to stay ahead of the game, to catch up, and to not be left behind.

An example is Cesar, a journalist in attendance at the Los Angeles hackathon, who worked on an interactive map that allowed users to see which areas of Los Angeles County had the highest and lowest rates of diabetes and obesity among Latinxs, as well as which areas had the highest and lowest life expectancies among Latinxs. He commented about the event: "Como periodista *no puedes dejar de avanzar*. Este hackathon nos permite explorar técnicas utilizando las últimas tecnologías y modelos que antes quizá no habíamos considerado." (As a journalist *you cannot stop advancing*. This hackathon allows us to explore techniques using the latest technologies and models that we hadn't considered.)

Fernando, a Chicago Migrahack participant, worked on creating an interactive web map that showed how undocumented people in detention in the US are moved around the country in a disorderly manner, often several times in just a few months. The visualization provides a dizzying experience where the overlapping of graph edges (representing migrant/detainee movements) connecting nodes (representing detention centers) over time

---

18. McLagan, 2012.

19. The minimal viable product requirement is more common in hackathons with a business orientation, where participants are asked to present a prototype that will satisfy "early adopters."

20. See Jones et al., 2015.

makes it almost impossible to follow any one specific journey.[21] Users can then click on individual detainees (with dehumanizing detainee designation numbers such as "19436") to learn about their personal stories. About his hackathon experience, Fernando states:

> We have to provide more opportunities like the Migrahack because they provide access to people and expertise. They create an environment, a very welcoming environment where one can explore, what for many people can be, intimidating. You know the world is moving at a very fast pace, *and if we don't catch up* . . . In fact it's not about catching up, we need to start leading.

These themes/fears of staying ahead of the game, of staying current, of not being left behind by technology, were frequent across interviews and in media portrayals of the events. Popular media reports in particular picked up on the diversity aspect of this hackathon, praising the organizers for putting together a structure that allowed those who would not normally show up to a hackathon to attend and become immersed in the code worlds; the reports praised the participants for taking it upon themselves to learn new skills in order to participate.

As the Migrahackers demonstrated, they were intimately concerned with what was being *made* or prototyped at the event. The storytelling was important, as was this idea of catching up. Like at other hackathons, the inherent time constraints at the Migrahack meant that many times the prototyped projects were never really made at all. At the Migrahack it was clear what was being *made* overstepped the boundaries of the projects at hand; what was being made were mindsets, hopes, futures, and participation models and subject positions to occupy these futures, as I explore in the next section.

[3] Participants-Who-Can-Participate

Each instance (to use a software design term) of the prototype, of the sociotechnical configuration manifested as an object to come, becomes a potential vision of a way of organizing society as a whole and the place of the "community" and individual within that society. The community aspect of prototyping was quite explicit at the Migrahack. As Armios gathered his

21. Many times private information like this is retrieved by submitting a Freedom of Information Act (FOIA) request. The FOIA offers the public the right to request access to records from any federal agency. At some of the Migrahack events, a workshop was offered to guide users on how to submit an FOIA request.

teammates at the Denver event to work on a project related to unaccompanied minors, for example, he asked each interested individual to introduce themselves by answering, "Tell us about the communities you're accountable to." This invited members to reflect on not only which communities they belonged to, but who *they* were, how they defined themselves in themselves terms of belonging to multiple, perhaps contradictory collectives; it asked of them to begin to prototype a coherent picture of who they were. Prototypes are, by definition, incomplete—they invite makers to work on completing the object at hand. In this case the invite to complete the object is coupled with an invite to complete their selves. Analogous with "design thinking" and "thinking with your hands," hackathon rules encourage participants to have fun, "break rules," in the process of creating new sociotechnical objects and new selves.

The prototypes that emerge at Migrahack are not only aimed at improving society—in this instance by approaching issues related to immigration and inequality—but are also aimed at constructing "a way of looking at the world in which individuals constantly remake themselves, in which they test themselves against the world, and, if they find themselves wanting, improve themselves."[22] From interviews and from participant-observation across the various events, it becomes clear that migrahackers come with genuine desires to improve themselves and society; it is also clear that they are interpellated as subjects who *want* to participate, who *want* to improve themselves.[23] They are not only performing their burgeoning *Latinx maker* status but also their ability to exercise their neoliberal subjectivities, to construct, mobilize, and manage their own Latinidad.

"Fail fast, fail often" is a common phrase that circulates in hackathons. The phrase indexes the fast-paced, disciplined risk-taking that is carefully honed at these events, and which (quite explicitly) mirrors Silicon Valley or "California" ideologies.[24] The "failure" of the prototypes (i.e., nothing becomes of the startups or projects beyond the hackathon) and the "failure" of the teams (i.e., they might just "shake hands and say goodbye" after the event) is expected not only by the organizers but also by the participants.[25]

22. Turner, 2016: 262.

23. Louis Althusser (1971) describes interpellation as the process whereby we become the subjects we are by responding to the hail of ideological formations that structure our social environment.

24. While other technological global "nodes" (e.g., "Silicon Alley," "Silicon Valle," "Silicon Savannah" [see Poggiali, 2016], or techno-capital hubs in Israel or India) are sometimes referenced to compare infrastructures, the model point of comparison is undoubtedly California's Silicon Valley.

25. See also Irani, 2015.

Likewise, participants in the Migrahack formation do show up with visions for advocacy and future calls to action—an ethos that mirrors the discourse of the event organizers. "The results: Apps, stories, graphics, maps—and friendships. It's powerful. It works. With training and mentoring, open data is an opportunity for all," states a promotional video for the event. But the hackathon participants don't necessarily expect that their apps will be completed or that their budding friendships will last too long.[26] Indeed, the only feasible way they could "fail" is by not being at the event, by not becoming participants, and by not taking advantage of the possibilities that this opportunity to *participate* presents.

Scholars of the "participatory turn" argue that participation has evolved into a leading mode of subjective interpellation in our contemporary period. Participation is construed as not only a concept and a set of practices, but as "the promise and expectation that one can be actively involved with others in decision-making processes that affect the evolution of social bonds, communities, systems of knowledge, and organizations, as well as politics and culture."[27] Especially with new media technologies that purport to create egalitarian technical infrastructures and modes of engagement where everyone can participate, participation becomes desired, expected, and ultimately, normal.[28] Participation is "often understood as a *problem*: How to get more (or less) participation? How to improve the quality of participation? What motivates people to participate (or not)?"[29] Not to participate is seen as strange and disappointing; the non-participant becomes suspect.

As Migrahack attendees build their prototypes over the weekend, they fulfill the promise and the expectation of participation: that one can be actively involved with others in decision-making processes that influence the construction of social bonds, "communities," systems of knowledge, organizations, politics, and culture. If the construction of a class of active citizens is formulated in relation to a class of excluded citizens, *participants-who-cannot-participate,* Migrahack attendees avoid their structural exclusion by materializing their subject positions as

---

26. The hackathon itself can function as a space where participants might be able to network to find work. For a full discussion of the different strategies of networking to find employment, and the relationship of these strategies to the global (neoliberal) economy, see Gershon, 2017.

27. Barney et al., 2016: x.

28. Fish et al. (2011) provide a "birder's handbook" to the forms of participation and the range of theories used to understand participation, from "peer production" to "presumption" to "networked publics" to "user-led innovation."

29. Kelty, 2019: 4.

participants-who-wish-to-participate, participants-who-decide-to-participate, and *participants-who-can-participate.*[30]

In order to claim a place in this latter category of active technical citizens, of participants-who-*can*-participate, hackathon attendees must not only be able to replicate the discourse of the event—and clearly many were able to do this, as can be seen by examining the quotes I have introduced—but they must also *perform* their technical understanding and capability, their *hacker* ethos and *hacking* abilities.[31] As the workshop day comes to a close at the Mexico hackathon, Cesar enthusiastically tells me,

> "Ahora programar no se me hace *tan* intimidante. Se trata de reusar cosas y a veces alguien ya programó lo difícil por tí. No tienes que entender todo para agregarle una nueva función al programa." (Now programming doesn't seem *that* intimidating. It's all about reusing stuff and sometimes someone has already programmed the hard part for you. You don't have to understand everything to add a new function to the program.)

Cesar, like other participants at the event, have picked up on what it means to navigate the layers of abstraction of the computing stack, and in particular, the benefits of an implementation guided by loose coupling, where you don't have to know the implementation of a particular component to add a function to it. What was more *tightly coupled* was his deep knowledge and care for issues related to Latinx and Mexican migrant health issues and life expectancy rates, connected to issues of healthy food access, gentrification, and access to quality healthcare. The accountability to his community and these issues that affected people he cared about deeply were inextricable from the way he learned to navigate this instance of the ethno-stack, inseparable from the way he experienced the hackathon. One of the most obscure components of participation, least studied and least taken seriously, is the *experience* of participation.[32] This experience is fundamental to how a participant becomes part of a collective. The procedures of participation thus become ways of "making up people."[33]

---

30. Barney et al. (2016) use the term participants-who-cannot-participate to reference the material reality of a class of citizens present in Aristotle's classic formulation of citizenship, whereby slaves and women "belonged" to the household and were excluded from "the administration of justice and the holding of office" *as a condition of the possibility of participation* by Greek male citizens (italics in the original: x). Kelty (2019: 15) asserts that any practical definition of participation implies it is voluntary, that an individual has to *decide to participate.*

31. Uribe (2021: 185) further explores how Mexican hardware hackers replicate the discourse of a state-sponsored Maker Faire, recirculating narratives that frame them as "voluntary makers."

32. Kelty, 2019: 3.

33. Kelty, 2019: 16.

In the next section I ask, how is it that the *code work* taking place at the Migrahack, aimed at constructing prototypes, relates to the prototyping of a particular people?

[4] Prototyping Latinidad at the Migrahack

The hype around hackathons aimed at empowering particular groups of people is based on these collectives not only learning to use technology but also learning to actively participate in building the technology that they use. Regardless of what level of the computing stack Migrahack participants found themselves in, or the expert or novice designation they aligned themselves with, one thing was certain: they were not only becoming makers but *Latinx* makers. The construction of the "Latino" racial/ethnic/identity category has been studied by various scholars who straddle disciplinary boundaries with Latinx Studies.

Cristina Mora, for example, examines the rise of the "Hispanic" category by examining what she calls the "politics of categorization" in the US between 1960 and 1990.[34] Using archival work, she looks closely at documents from organizations like the National Council of La Raza (NCLR). How was it, she asks, that by 1990 the NCLR had transitioned its agenda from one focused on protest activities (protesting local instances of discrimination against Chicanos and Chicanas) and regional programs (funding and implementation of job training programs and daycare centers) to a national agenda focused on policy analysis and research, legislative advocacy, and lobbying focused on a national Hispanic constituency?[35] A lot of this transition was based on the financial pressure placed on the NCLR to become the nation's foremost Hispanic civil rights advocacy organization—in order to gain power, the organization needed to secure resources from federal agencies, philanthropic foundations, and corporate foundations. The best way to do this was to frame its constituency as a national organization that was unified by *something*—in this case a hazily defined set of cultural values and experiences; these fuzzy values and experiences would come to shape a "Hispanic" community/group/category.

Mora thus points to the state-activist-media networks where diverse subjects could work together because, while they framed Hispanic pan ethnicity differently, they also referred to this common, albeit ambiguous, narrative

34. Mora, 2014: xii.
35. Mora, 2014: 51.

about Hispanic cultural values.[36] In the process, they became reliant on one another for expertise, data, and resources. "In the Hispanic case, activists became Hispanic political analysts, census officials became Hispanic data analysts, and media executives became Hispanic marketers."[37] There were ongoing negotiations between several sets of actors, each of whom had distinct interests and abided by distinct organizational logics. What kept the network together was the elision used in defining the new Hispanic field. They "never really defined who Hispanics were, nor did they argue definitively that characteristics such as language, place of birth, or surname made Hispanics Hispanic."[38] Participants in these networks used descriptors like "hardworking," "religious," and "family-oriented" (adjectives that could be used to describe any group) to give Hispanics a common set of values and a common "culture."[39] As Cristina Beltrán argues, Latino or Hispanic categories are "terms whose descriptive legitimacy is premised on a startling lack of specificity."[40] The "inclusivity is part of the problem: 'Hispanic' and 'Latino' tell us nothing about country of origin, gender, citizenship status, economic class, or length of residence in the United States."[41] The terms are even racially indeterminate, as Latinos might be white, black, indigenous, or any combination of these. Defining someone as "Latino" or "Hispanic" becomes "an exercise in opacity—the terms are so comprehensive that their explanatory power is limited."[42] While categories that Latinos/Hispanics come to occupy (or identify with) are constructed at an institutional level, from above in a sense, scholars have shown how subjects who identify with the Latina/o or Latinx categories—whether they are embedded in these decision-making spaces/institutions or not—differentially perform and contest these descriptors, many times creating "horizontal hierarchies," intra-Latinx conflicts and tensions.[43]

But the comprehensiveness of the categories also has its benefits for those who wish to wrap themselves in its inclusivity, as Latinidad provides opportunities for solidarity-making. Latinidad can thus become "the condition of being Latina/o," anchored in the "social, everyday realities of our

36. Mora, 2014: 6.
37. Mora, 2014: 13.
38. Mora, 2014: 156.
39. Mora, 2014: 156.
40. Beltrán, 2010: 6.
41. Beltrán, 2010: 6.
42. Beltrán, 2010: 6.
43. Aparicio, 2019; Dávila, 2001; De Genova and Ramos-Zayas, 2003.

diasporic communities and in the spaces populated by Latinas/os of various nationalities, generations, immigrant statuses, and racial and gender identities."[44] The term makes a morphological shift "from a label of identity to a doing, a political and liberatory action."[45] This call of Latinidad is taken seriously by Migrahack participants, in that the horizontal hierarchies and tensions seem to be minimal, and that they've made this shift to rethinking Latinidad as a form of *doing*. The space seems to have been carved out from the "real world" as an experimental laboratory where hopes, futures, and identities are carefully proposed and inspected beyond the reach of those usually in power. Working a different angle to examine how a Latinx space of solidarity-making and community organizing emerges, Inés Casillas explores how (mostly working-class) Latina/os carefully construct a space where they not only get to "be themselves" but also where they are able to engage with each other in order to deliberate their position within a broader national body.[46]

Specifically, Casillas shows how radio has become the medium of choice for these community negotiations.[47] That is, "as an aural stage, Spanish-language radio provides Latinos the opportunity to retreat and deliberate outside the surveillance of dominant society, and engage emotionally and economically with more than one national body by also affirming their distinct class and ethnic identities."[48] She argues that "at a time when visuality overwhelms most media formats (film, movies, television), sound offers a unique platform for a listenership that is characterized by language, class, mobility, and, for many, legal status."[49] Casillas makes the point that these question-answer sessions on the radio are often aired live, and that these dialogues carry elements of an oral tradition long familiar to Mexican and Chicano communities. Most importantly, as audiences deliberate and organize around critical transnational issues for US Latinos—listening offers an opportunity to retreat but also a sense of anonymity for collectives dependent on inconspicuous livelihoods.

44. Aparicio, 2019: 113.
45. Aparicio, 2019: 117.
46. Casillas, 2014.
47. The Federal Communications Commission (FCC), which functions as official watchdog for radio broadcasting, is unable to keep up with monitoring Spanish-language radio. This is for a simple reason: they lack bilingual investigators. In 2004, they only had two Spanish-speaking investigators on staff. Thus, institutional racism in this case plays in Latina/os' favor, as a lack of monitoring makes transgressive sound practices possible (Casillas, 2014: 10).
48. Casillas, 2014: 9.
49. Casillas, 2014: 9.

Similar to this self-organized Latinx radio community, Migrahack participants are able to think through issues relevant to their communities without
worrying about outside monitoring or judgment. *Or are they*?

At one of the US-based Migrahacks, a set of snapshots from the event
might tell us otherwise. After the teams formed, they delegated tasks to
different team members, working diligently against the time constraints of
the weekend event, and spreading out across the college campus where the
event was held. Some teams found private conference-style rooms; others
found shade at outdoor picnic tables. But one group of subjects stayed in
the main room where the opening remarks of the event took place. Several
people here took it on themselves to run a sort of central command post
from this space. One of their tasks was finding teams for people who arrived
later in the day to the event. As a new potential hacker would arrive, the
central command helper might do a quick interview of the person to assess
their skills and interests, and quickly integrate them into a team. During the
second day of hacking, while teams were especially immersed in their hack
modes, manipulating datasets, experimenting with web visualization tools,
conducting impromptu research interviews, and planning out their pitch
sessions, I stopped by the command center to get a feel for the type of work
going on there. The ambiance was markedly less intense. This was probably
because there was far less work going on there. Folks were laughing out loud
and discussing things not at all related to the Migrahack. The demograph-
ics were also markedly less Latinx. This group was overwhelmingly white. I
wondered why some of the people in here had not integrated themselves
into some of the teams that could have used the extra help.

I wasn't the only person that noticed these differences or that wondered
the same thing. I found Armios resting in a comfortable egg chair in a lounge
outside of the library, looking a bit depleted and mumbling something to
himself. "Can you believe they even want us to be their *research participants*?"
he told me, pronouncing "research participants" in a mocking, "official"
sounding tone. He showed me a lengthy consent form with an attached
survey. A researcher from the university had designed and handed out the
15-page survey to study the "social and political attitudes" at the Migrahack
event. Armios didn't fill out the survey. Even if he wanted to, he might not
have had the time to give it his full attention, given the workload and learning
curve participating at the event entailed for most participants. But Armios
seemed to have refused to fully engage in the event; he left his team even
though he had initially emerged as a potential leader, and I didn't see him at
the final pitch session. His remark to me about the survey, and his pointing

to this particular *they*, different from him and different from us, made me think that perhaps he had grown annoyed with the power differentials he had witnessed, or more likely, experienced. Perhaps the heavy burden of being a "research subject" for those working out of central command was not what he had signed up for.

If the experience of participation is a key component of how people form into collectives, participation is also fundamentally about *experiencing power*. Armios felt the power of surveillance, the power of inequal work dynamics, the power of the gaze. The experience of participating can be one of perplexity, where participants suddenly become aware of a background or a social dynamic that becomes exacerbated in the act of participating.[50] And participating can also entail deciding *not* to participate. Hacking usually connotes a countercultural ethos, and if one aligns with the hacker ethic, it means you break rules and " <2> Don't ask for permission." As Migrahack participants are called upon, and take seriously the call to "think with their hands," as well as to construct Latinidad as a type of *doing*, I've argued that this *making* also becomes a way to think through the *making* of themselves as a people. The Latinx makers become responsible for making their own Latinidad; as they prototype their projects, they prototype a new way of embodying these newly forming hopes, stances, and futures, as well as the subtle power relations and social dynamics that emerge at the event. Migrahack attendees are invited to become "participatory," "collaborative," "engaged," "concerned"—and of course "hackers" and "makers"—but by whom, under whose watch, for what reasons, all become questions that inherently emerge from a collective used to nebulous and shifting markers of racialized identity, based on fuzzy values and descriptors, that become even more complicated when communities from both sides of the US/México border work together.

[5] Always Already and Never Quite Yet

The iterative ways of the prototype, a critical component of software development, invites participants at the Migrahack to become one with the idea of incompleteness, of developing versions that might become better iterations in some future. If prototypes are by definition incomplete, how do prototypes contribute to Latinxs' perceptions about their own incompleteness? This is especially important when considering that "Latinidad, in its joint

---

50. Kelty, 2019: 20.

articulation alongside prevailing forms of racialized difference, including Africanness, Asianness, and various American Indigeneities, is more than 500 years in the making, yet always on the demographic horizon."[51] As Jonathan Rosa argues, Latinxs occupy a peculiar "social tense of the always already and never quite yet."[52]

When the US/México border is presented as something to be hacked, it takes center stage and makes its presence fully felt. Rihan Yeh's ethnography with communities who reside in the borderlands, next to the physical border, reveals that ambivalence and middleness emerge as issues again and again in their everyday lives, haunting the "everyday details through which people hope to claim modernity."[53] "As the physical scar slashed by the 1848 excision of almost half the national territory, the border is at the heart of the problem of selfhood and subjectivity, passing and prohibition posed by Mexico's unequal and denied colonial relationship with the United States."[54] Likewise, at the Migrahack, hackers from both sides of the border are confronted with the border, asked to hack it but also to take responsibility for their own becomings, for their own passages into "modernity"—they're tasked with resolving their own incompleteness and middleness by connecting practices of coding with constructions of selves and collectives as they negotiate their sociopolitical reality.

In this chapter, by exploring the role that prototypes and participation play in the renewal and completion of a people, and by thinking with the experts who form US/México transnational alliances in order to work on issues relevant to a dynamic Latinx collective, I've demonstrated what else is being made while they're making, what else is being coded while they're coding. As they align themselves with different layers of the computing stack, they make clear that their instance of the ethno-stack is tightly coupled, or as I've shown, passionately or intimately coupled. That is, the personal and interpersonal is not abstracted away as they work on the sociopolitical layer that the event sets out to hack. And with the permeating presence of the concept of iteration, connected to the prototyping, they make traversals across this ethno-stack by thinking with the sociotechnical layer as well. The construction of a "people" here is an elegant coordination between the experiences of individuals with a technology, as well as the experiences

51. Rosa, 2019: 15.
52. Rosa, 2019: 15.
53. Yeh, 2018: 17.
54. Yeh, 2018: 17.

of individuals to the degree they feel they are (or aren't) participants in a broader collective.

Migrahack's mission to bring the power of technology together with different "community" members in order to resolve complex transnational immigration issues is quite powerful. Participants are invited to learn to open up the black boxes of technology and in the process learn about the code work it takes to move between them. It gives them different tools to think about how the politics of migration "work." Although the events are not marked specifically as "for Latinxs," many of the participants identify as Latinx and many of them, as the advertising for the event states, have not interacted before. If scholars have identified advertising, radio, and language as domains where Latinidad is constructed (and contested) when Latinxs are asked to become *Latinx* cultural producers, here I argue that "hacking" might be another domain where Latinidad is constructed and negotiated, in much more subtle ways. Relations of power and privilege are foregrounded at different moments during the hackathon: who's putting in the work and who isn't, who from Mexico has the privilege to cross the border to attend the US events, who gets to make a researcher's career out of "analyzing" the way Latinxs resolve their problems. The unequal relationships and corresponding subjectivities are experienced but also cultivated within these spaces that form the techno-Borderlands, and they eventually spill out into domains beyond the code worlds and the hacker worlds.

As particular iterations or versions of Latinidad come to be objectified and mobilized strategically across borders, collectives shift from thinking about hacking migration to considering how perhaps migration itself might be a way to hack. I continue this line of thinking in the final chapter by focusing on how research participants think with "the pivot," a tech startup term that calls for changes to a product that might better align it with the market, to manage and perform their Latinidad across national, racial, and ideological lines.

[6] Pivoting across the Techno-Borderlands

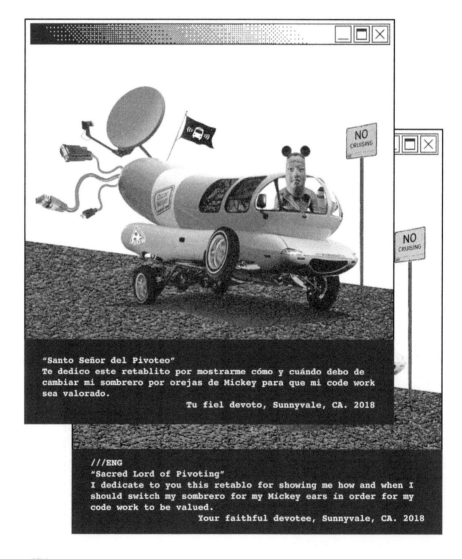

"Santo Señor del Pivoteo"
Te dedico este retablito por mostrarme cómo y cuándo debo de cambiar mi sombrero por orejas de Mickey para que mi code work sea valorado.

Tu fiel devoto, Sunnyvale, CA. 2018

///ENG
"Sacred Lord of Pivoting"
I dedicate to you this retablo for showing me how and when I should switch my sombrero for my Mickey ears in order for my code work to be valued.

Your faithful devotee, Sunnyvale, CA. 2018

[0] From Politics to Pizzas

"These entrepreneurs are bright, well educated, hard-working, and excited—you can see the sparkle in their eyes." Saeed Amidi introduced the entrepreneurs that would pitch their startup ideas to the investors gathered at Plug and Play Tech Center in Sunnyvale, CA, better known as "Silicon Valley in a Box" to various tech startup collectives. The event was publicized as "Mexico Day"; 10 startup teams that had already gone through accelerator programs in Mexico would try to convince San Francisco–based investors to invest anywhere from $30,000 to $250,000 USD to move their operations to the Bay Area. "Why Mexico?" Saeed asked. "Why not—it's right around the corner, and it's easier to fly to than Russia," he laughed.

Saeed's joke was meant to foment the convivial ambiance; we were all encouraged to "connect" and make new friends and potential collaborators at the event. In case we needed reminders of the transnational collaborations being materialized, US and Mexican flags waved side by side as the center-pieces of every table, and the same giant flags were prominently displayed on the center stage. Saeed went on to give some background on Plug and Play. This was *the* place where Silicon Valley connections were made. From bottled water to the latest online workspace platform, this is where venture capitalists, investors, university professor-advisors, and the "sparkly eyed" entrepreneurs had come to connect and make million-dollar deals. This is where the hackers met the entrepreneurs to "let products grow naturally," and where everyone was able to become part of the journey. "Almost like a movie," Saeed cheerfully exclaimed.

He might have been referring to *The Social Network,* the movie that tells the origin story of Facebook. Walking into the room where the event was being held, you might have felt as if you were in fact a part of this movie, if only by the appearance of the people in attendance. Of the approximately eighty people at the event, there were about thirty "hungry entrepreneurs" (the Mexicans presenting their startups) who could pass for the Mark Zuckerbergs of the Silicon Valley. They resembled the popular tech mogul not only because of their light skin color but also because of their "hacker" attire—jeans, tennis shoes, "scruffy" overall appearance.[1] If it wasn't for the flags and the event being called "Mexico Day," I could have certainly been fooled.

Many of the entrepreneurs, ranging from early twenties to mid-thirties, only one of them a woman, came from prestigious private universities in Mexico (most from ITESM, Instituto Tecnológico y de Estudios Superiores de Monterrey, "the Mexican MIT"). Some had degrees from prestigious US universities such as Brown University and Harvard University, others had previous experience holding leadership positions for large corporations in Mexico, and some founders revealed that they already had substantial investment from "friends and family." But not all of these young entrepreneurs were elite, privileged Mexicans looking to increase their wealth.

Take the story of Javo, an amiable young man in his mid-twenties. He, along with his co-founder at the time, told me about their original idea, which was to use Twitter to report corruption in Mexico, namely, voting poll fraud during presidential elections. In contrast to reports that proliferated in 2018 that "malicious coders" promised to cause havoc in US elections by exploiting vulnerabilities in technical infrastructures,[2] Javo and his co-founder aimed to flip this narrative in Mexico, and instead use hacking to *prevent* fraud. The infrastructure they had designed was aimed at interfering with common tactics that political parties used to skew voters and voting counts. Politicians were known to not only "buy" votes from citizens by offering them money, debit cards, or even household appliances but also organize these voters (usually from marginalized populations) and bus them to strategic locations to cast their vote. Javo and his co-founder's app would use Twitter to help a crowd-sourced team quickly and efficiently arrive at locations when suspicious buses and crowds arrived at these locations.

---

1. AnnaLee Saxenian (1996) argues that this "laid back" California attitude, complete with corresponding wardrobe, in direct contrast to the "buttoned up" style of the East Coast, was an important determinant of Silicon Valley economic success. This "culture" led to more openness and cooperation across companies. Similarly, Aihwa Ong (2006: 168) finds this new "casualized work ethic" across Asia, where "floppy hair, jeans, and rolled-up sleeves" is linked to "the can-do, technologically savvy, and entrepreneurial figures celebrated in American neoliberalism."

2. This article, Giles (2018), for example, warns the public of how hackers might "cause havoc" in the 2018 US midterm elections. Pointing out that the current voting infrastructures contain outdated machines and operating systems, and that local networks have been set up by technological amateurs, the article convinces the reader that these "malicious coders" are in a position to exploit these technical vulnerabilities. The "worrying precedent" here is that the US Department of Homeland Security reported that Russian hackers had meddled in the 2016 US presidential elections, scanning computers and networks for security codes.

They could possibly interfere or at least document the occurrences in order to build cases against the culprits.

But after meeting with more seasoned startup advisors, Javo "pivoted" their project to Twitt2Go. "Pivoting" was a widely circulated buzzword in the tech startup community. Many traced its origin to *The Lean Startup*, a popular book that promoted protocols for efficiently developing tech products to meet the needs of early customers, and specifically proposed the pivot as a strategy for working flexibility into your project idea in order to be able to change it quickly to something that stuck with users, or in many cases, with investors. This new mobile application, Twitt2Go, would allow users to use Twitter to easily order food, in particular pizzas, from their favorite restaurants; this idea fit the needs of the market more closely and would probably catch the attention of some of the investors at Plug and Play in Sunnyvale at "Mexico Day." The technical infrastructure would work in a similar way to their voting poll fraud system, but would simply organize delivery persons and pizza consumers instead of voters and politicians. Their original idea might have been *pivoted*, or better yet, distorted beyond recognition, from politics to pizzas (more marketable products), but at least it gave them an in to brush shoulders with Silicon Valley and Bay Area entrepreneurs who might help them get their startup to the next phases of development.

Curiously, and against the stereotypes of Mexican migrants, Javo and his co-founder's ultimate goal was not to stay in the US but to return to Mexico; they fell neatly into the category of highly qualified migrant workers looking to contribute to "brain circulation" instead of "brain drain."[3] Moreover, they told me that if they made enough money, they might be able to move back to Mexico and continue working on the social issues they were passionate about in the first place, before the pivot. "Para tener impacto, necesitas feria" (To make an impact, you need money/bling), Javo tells me, confirming his long-term vision.

Those who controlled the *feria* at the event were the investors, older men mostly wearing suits to distinguish themselves from the young entrepreneurs, who listened attentively and whispered remarks to each other as the (mostly) young Mexican men presented their startup ideas. Joining Javo, who presented Twitt2Go, were the founders of PingStamp, who promised

---

3. Saxenian, 2006; Tigau, 2013. For a sociological perspective on the recruitment and incorporation of "high tech braceros" from Mexico, see Alarcón, 1999.

to bring purchase loyalty programs to Mexico; PaydayLoans, who pitched a mobile app that would allow users to receive a personal loan within 15 minutes; and other hopeful founders who took turns pitching their ideas on the stage.

After the presentations the investors had the chance to ask questions and give advice. A common question emerged: "What is a similar application that has been successful in the US, and how can you do the same thing faster and more effectively in Mexico?" "Find something that worked in the US, and execute the hell out of it in Mexico," another investor said. Saeed chimed in, "We need at least two Googles and Dropboxes in Mexico." After PaydayLoans presented, one investor offered advice about their presentation: "Move your advisors to the very front. I was nervous and skeptical about your startup until you showed me who was behind it." If any of the young Mexican entrepreneurs didn't know it already, at "Mexico Day" they learned that "pivoting" your idea in the tech startup world was inextricable from the circulation of economic capital as well as classed cultural capital, and perhaps more importantly, that to reap the benefits of these forms of capital, pivoting your startup also meant strategically pivoting other elements of your presentation of self.

In this chapter I demonstrate how hacker-entrepreneurs negotiate their sociopolitical and economic reality by thinking with the pivot. They pivot their identities, their language practices, their presence and presentation of self as they work across all layers of the ethno-stack, the personal, interpersonal, sociopolitical, and sociotechnical. As the comments from the investors and the participants at Mexico Day demonstrate, their alignments across this ethno-stack become even more complex when we add nationalized and classed borders that call for creative traversals across its layers. I show how Mexican and Latinx hacker-entrepeneurs reconfigure the market logics of agility, competitiveness, and risk to creatively combine them with logics of hacking characterized by reinvention, playfulness, and resistance. They move and think with the pivot to manage and perform their Latinidad and labor potential across nationalized and racialized lines, sometimes brushing up against each other's pivots. I follow the trajectory of Javo closely across the techno-borderlands to show how his app ends up returning to politics in surprising ways, arguing that this happens because his team not only mobilizes migration as a type of hack, but also focuses their code work at the root layers of the ethno-stack.

[1] Sombrero-ed Coders OR Coded Sombreros?

This wasn't the first time Javo had paraded with Mexican flags in front of an audience. The previous incident occurred as part of an annual software developer festival sponsored by a large, popular tech company in 2015. These events were becoming increasingly widespread and obligatory for tech companies to be considered among the top tier of tech giants. The festivals featured talks, demonstrations, releases of new products and product features, and, in the latest iterations I attended during fieldwork, guest celebrity speakers. Some of my research participants from Mexico would use their savings to make the annual pilgrimage to the headquarters of their favorite tech company in Silicon Valley or in San Francisco. Luckily for those who couldn't travel in person to the 2015 event, this tech company sponsored "extended" versions of the main event; these additional physical locations across the world featured contextually and geographically relevant live presentations and events.

As the live feed from the annual festival in Silicon Valley was streamed to these extended node locations, so too were video streams of these locations broadcast back to the headquarters in the Bay Area. As the keynote speaker began his address, he wanted to show images of these locations to all viewers. This practice might sound familiar from events such as globally televised sporting events, the World Cup, for example, where distributed collectives view the event at the same time to feel the solidarity and collective effervescence of rooting for their team.[4] In this case, the team to root for was definitely the tech company, as was made clear when the keynote speaker started with the live feed from Mexico, much to the delight of the crowd in attendance in Mexico: Javo and hacker-entrepreneur friends packed into the auditorium and performed their overwhelming excitement on screen.

They all stood to their feet, waved their arms in the air, and screamed at the top of their lungs, first with random shouting and hollering and then, as if previously planned, into a coordinated, "Mexico, Mexico, Mexico!" chant. If an outsider were to see the crowd without knowing that they were watching a tech company's live transmission, they would likely conclude that this crowd of mostly young men was watching the Mexican national team score a goal in the World Cup. If not for the excitement and chanting in the name of their country, then the reminders would have been the sombreros, the waving

4. See Joo, 2012.

Mexican flags, or the clothing that coordinated with colors of the Mexican flag. The one young man who showed up in a mariachi suit would have been the dead giveaway.

Javo was not the young man in the mariachi suit, but it was still somewhat surprising to see him participating in these performances that worked to essentialize "Mexicanness." That is, in previous interviews with Javo, he had told me about the pushback his startup team had encountered when trying to raise capital for their projects, in their various iterations: the voting fraud infrastructure, its "pivoted" pizza delivery system, and a third project, an app that used mesh networking to help mobile phones connect with each other without wireless internet or telephone signal. US-based investors met his team's proposals with questions about his team members' origins, "Are all the founders Mexican?" for example, and doubts about the way they would run their operations: "You're all Mexican so you will be running things in the 'Mexican way.'" Javo resented these generalizations and felt that they were detrimental to all the code work and dedication he and his team had already put into their projects. He told me that before every meeting he always felt nervous that potential investors would first consider their national identities and essentialize them and their projects without carefully considering the content and value (both economic and social) of their ideas.

Thus, after getting to know Javo and his experiences as well as his reflections about these encounters in both the US and Mexico, it initially surprised me that he was self-essentializing his self in front of a live feed, performing the stereotypical "Mexicanness" that had so explicitly prevented him from accessing the necessary capital to carry out projects he so deeply cared about. But actually, upon further reflection, his performance of the stereotypical Mexican at the event made sense. If he harbored hard feelings from the encounters with US investors and entrepreneurs, the tech company's extended event in Mexico functioned as a type of safe space from which he could perform his Mexicanness without being judged or put down. The exaggerated displays of collective identity and nationalist pride in effect served as a symbolic response to the rigid constraints on cultural and nationalist identity that the US investors seemed to unwelcomingly impose on him. But how to transform these responses to have material and economic effects? To further understand his motives and his corresponding moves across the techno-borderlands, we need to trace how Javo's code work aligned with concepts such as "the pivot" to make moves that countered the flexibility and the "pivots" coming from the other side—those performed by the companies and the investors themselves.

[2] Pivoting Presence

Javo and hacker-entrepreneur friends at the annual tech company festival effectively performed their Mexican *hackerness* by participating in the event; they also strategically performed their *Mexican* hackerness by making visible the stereotypical sombreros and other "Mexican" material artifacts. William Mazzarella finds similar underlying dynamics when a large multinational corporation deploys its strategies to expand to "global consumers," in his case the "Indian consumer."[5] Mazzarella states that the Indian consumer shares the structural doubleness that is characteristic of the commodity form in general. For the advertising and marketing professionals he conducts research with, the "Indian consumer" should have not only a general level of equivalence, the Indian *consumer*, the commodity that can be sold to multinational clients, but also an irreducible particularity, the *Indian* consumer, its corollary form that these advertising and marketing professionals can claim exclusive rights to, in terms of production and distribution.[6] These Indian "versions" meant that there needed to be mediating professionals who doubled as cultural guides in order to give the corporations access to consumers who were "hungry to consume."

Curiously, Javo uses similar language to describe his participation in events such as the Plug and Play demo from the chapter's introduction. "You have to know when to leverage the fact that you're a hungry Mexican who knows the market, but you have to know who to tell and who not to tell that to. You can't tell a top investor in Silicon Valley, 'Hey we're based in Mexico and we're all Mexican,'" Javo tells me.[7] He has to learn when to deploy his Mexicanness and when to deploy his hacker-entrepreneurness. In the US, he tried to perform the latter, despite attempts by investors to pin him as the former; at the event in Mexico City, he resorts to displaying the former, since he was already being interpellated as the latter with the inclusion in the tech company's presentation. Javo learns to pivot between these two roles, attempting to perform the Mexican *hacker-entrepreneur* and the *Mexican* hacker-entrepreneur, to be effectively grounded as the "Mexican hacker-entrepreneur" who can navigate the flexible dynamics of the market but also provide access to the "talent" and "opportunity" his flexibility gives him access to. But to whose benefit?

5. Mazzarella, 2005.
6. Mazzarella, 2005: 233.
7. Original quote in English.

From the perspective of the tech company that hosted the annual festival, their live video feed that flashed various geographical locations is a way to make their own flexible global "presence" visible in a system that asks employees (actual and potential) to perform the same. Just like Saeed from Plug and Play in the opening of this chapter compares Mexico and Russia, as places he and his team of investors can easily tap into for "talent" and "sparkly-eyed entrepreneurs,"[8] the tech company demonstrates across its feed where it has its eye and its reach. It's common to hear how large of an "untapped market" Mexico possesses, but the feed images also function to demonstrate the willing and able coders who also double as prepared cultural guides in each country; these subjects represent the tech company's *presence* in these locations.

In her ethnographic work with Wall Street investment employees, Karen Ho finds the surprising side of corporate "presence."[9] Her respondents confirm that they are sometimes part of a "team" of 1–2 employees in an international office where banks have an international location; moreover, an empty office also qualifies as a location and, more importantly, another notch en route to the company's globalness. The banks develop a "global strategy" under which they decide which location justifies an important and strategic market, but at the same time project the sense that they can be present in many (and any) markets with flexibility. One of the executives at the banks demonstrates his flexibility and presence when he says, "We do China; I like India,"[10] in the same way that Saeed at Plug and Play makes on-the-fly decisions about his group's decision to look for "talent" in Mexico over Russia. The comparisons of geographic markets, the live feed at the tech company's annual coder festival, and the empty offices that are intended to perform the "global presence" of Wall Street companies all point to "specific ways of constructing and imagining scale and movement in order to achieve particular goals and positions in a world of demanding financial flexibility."[11]

Taking into account these complex dynamics, we can better understand Javo's own performances in sites across the US/México border. Javo plays

8. Analysts and consulting companies increasingly framed Mexico as a place where there was a surplus of "talent" waiting to be hired for 2–3 times cheaper than the US, with specific geographical advantages over India. See, for example, Parés (2021) and a 2019 report by CodersLink, an organization dedicated specifically to connecting global companies with "world-class talent, no matter where they are located," which can be found here: https://coderslink.com/company/tech-salaries-report-2019/.

9. Ho, 2009.

10. Ho, 2009: 325.

11. Ho, 2009: 314.

with his own *visibility* across these borderlands. Among the reasons that investors gave him for not investing in his projects had been: that he wasn't raised in the US, so he didn't understand how the business world really worked; that his team was not based in the Bay Area, so it would be hard for investors to keep an eye on his operations; and that since he was Mexican, he would proceed to do things the "Mexican way." In an attempt to chip away at the refusals and become more legible to the investors, thereby making his companies potentially more desirable investments, he presented his startup companies as based out of the US. One way to do this was to register your startup as a Delaware C Corporation. While it was technically possible to incorporate your startup as a US legal entity using the "C Corp" structuring in any state, Delaware offered specific tax benefits to investors and became known as a favorite for capital venture firms interested in early-stage tech startups.[12] In addition, other Mexican hacker-entrepreneurs circulated information in Mexico about how to register your startup as a Delaware C Corp, making it less intimidating for others to submit the relevant legal and financial documents. Thus, by Javo officially registering his startups as "American companies," he could complete his performance of presence with a San Francisco address and present all of his app's documentation in English—from an outsider's perspective, it was a US tech startup company.

Indeed, across the hackathon and co-working spaces I participated in, teams would consistently give their startup projects an English name as a first step to performing their companies' US "presence." At one hacker-space in Mexico City, for example, I worked with Memo and two of his hacker-entrepreneur friends who implemented an augmented reality app that would present users with offers from local street vendors as they walked through specific physical locations. Their first option for naming the app was "DoñaCholeApp," which used the name Doña Chole (Mrs. Chole) as a generic reference to the (mostly) women who sold tamales, esquites, tacos, tlayudas, pozole, and other prepared food items on the street. The name also reflected the aim of the app, which was to incorporate these street vendors (the "Doña Choles") into tech platforms that usually forget about them. Their second option was "MerolicApp," a composite of *merolica,* a

---

12. The "C" in C Corporation refers to the corporation's tax designation. A Delaware corporation taxed as a C Corp is often preferred and well adapted to venture capital and angel investing over other forms of business entities because they are "founder-friendly." In addition to providing limited liability to the owners, they frequently have financial benefits such as allowing for 10,000,000 shares of common stock authorized with $0.0001 par value, or issuing half of the shares to founders.

Mexican slang term used to describe a street vendor who attracts unexpecting transients with their clever verbiage, and "app." They continued to play off of local slang that referenced street life and vending with Spanish names that I thought were especially creative, but ultimately, though, they ended up naming their app with a much more boring English name, "Offertunity." Even though the app aimed to address a very local problem meant to be used by mostly Spanish-speaking users, Memo and friends eventually became adamant about naming it in English. "Nos dimos cuenta que si piensan que somos una startup Mexicana, nadie nos va a pelar" (We realized that if they think we're a Mexican startup, nobody will pay attention to us), Memo told me. Many other hacker-entrepreneurs confirmed that unless they made this English pivot, users would simply not download their apps.

More importantly, if Mexican hacker-entrepreneurs wanted to launch their apps or other startup projects, international (and specifically US) investors were unlikely to provide them with any attention, and much less with capital, as Javo confirmed firsthand. In the same way that the companies learned to perform their globalness by setting up shop in places they could quickly pull out of, so too had Javo learned to perform his globalness by registering his startup in the US. "It's not a lie because technically our headquarters are in the US even though we physically aren't," Javo assured me.

Javo was learning to navigate the overarching processes of capitalism and the knowledge economy. The tech companies saw him as an untapped tech consumer and producer in Mexico, but also as an unexploited "talent" that knew how to code and could provide access to other exploitable consumers/producers. He takes up the discourse to make himself visible (and recognizable) as a Mexican hacker-entrepreneur, and he wears and removes the sombrero as he makes the necessary pivots of his company (and his self) between the US and Mexico. The sombrero works here as a material artifact to demonstrate his moves but also as a metaphor to the types of negotiations and alignments he makes to effect these transitions. In the next section, I show how markers of race further complicate the pivots that need to be made across the US/México techno-borderlands.

[3] "Perfect English" AND Latinx Frictions

Javo was one of the young men I followed closely as I conducted fieldwork across the techno-Borderlands. His moves between Mexico and the US and his strong construction of self as a "hacker-entrepreneur" reflected the flexibility, contradictions, and mobility that made research with Mexican hackers more revealing and complicated. Near the end of my fieldwork, Javo and other hacker-entrepreneurs I had developed close relationships with participated in an event that surely would have interested many of the audience members present in spaces where I have subsequently presented my research. That is, it would have interested audience members who consistently and inevitably asked the following question: "So what is *different* about hacking in Mexico?"

At a coder festival held at the ITESM, Javo and fellow travelers of the code worlds across the US and Mexico held a panel titled "Diferencias entre EEUU y México" (Differences between the US and Mexico). The small university room was packed to capacity, with young people and a few journalists in attendance. The young men took turns recounting their experiences attempting to launch startup companies in the US and Mexico, and effectively took the role of cultural "experts" who understood the cultural differences between the US and Mexico.

The comparisons took familiar routes, unfortunately. The panelists circulated stereotypes about each of the collectives up for comparison, those that represented the "Americans" and those that represented the "Mexicans." Americans in Silicon Valley, they claimed, were prone to sharing; Mexicans were not. Americans worked efficiently to avoid *trabajo doble* (double work); Mexicans were perpetually stuck *trabajando doble.* Americans were more direct and said "no" when they needed to; Mexicans were culturally unable to ever say "no" and were forever stuck replying "yes" to every request, endlessly accepting work they already knew they could not accomplish. Ultimately, the panelists circulated common narratives about the competitive cultural advantage of Silicon Valley[13] and spread developmentalist narratives that urged their fellow Mexicans to "change their culture," not dissimilar from the government's *todos con el mismo chip* initiative that promoted "English and computing" for everyone as a means to save the country's economy. In addition, their comments fell in line with neoliberalizing discourse about taking matters into one's own hands, not taking no for an answer, and thinking "globally" instead of staying stuck with local concerns that wouldn't help your project scale.

Esteban, the ever-present Chicano/Mexican-American at these annual coder festivals, asked Javo a question that for a moment seemed to challenge their "anything is possible" recommendations. "Do you think you've had success in San Francisco because you speak English so well?" he asked

---

13. See also Barbrook and Cameron (2009) and Turner (2006) on the "California ideology."

Javo directly. He asked his question in English; he knew Javo spoke English, and since the panelists had been code-switching (between English and Spanish) during the presentation, changing the conversation to English seemed acceptable. In fact, this code-switching was particularly common throughout the event. This linguistic version of the "pivot" paralleled the other cultural and identity pivots Javo and his hacker-entrepreneur friends had become accustomed to. More so than in the southern part of Mexico and even Mexico City, people in the north of Mexico were arguably more accustomed to this language code-switching or pivoting, in no small part because of their proximity to the United States. Javo, sticking to Spanish, responded, "El inglés ayuda pero sólo hablar español no es un impedimento" (English helps but speaking only Spanish is not an impediment). Javo not only stuck to Spanish but also to his "anything is possible" rhetoric, claiming that it was not necessary to speak English to (presumably) travel to the US and secure investments. This was in fact the "success" Esteban was referring to, since Javo, at this point, had secured angel investment for his latest startup company, Pingafy, a platform that used mesh networking to "connect the unconnected." Esteban, keen on pushing Javo further on the matter, and perhaps putting him on the spot in front of the audience, asked, "So how do you speak English so well?"

Javo hesitated for a moment, but then proceeded to provide full-disclosure about his ability to speak what most would call "unaccented" English. He told the audience that his mother had spoken English to him since he was a child, and that her English came from her grandmother, who was born on the other side of the border, in El Paso, Texas. "Mi abuela tenía 'perfect English,' mi mamá tiene 'perfect English.' Mi inglés suena mejor después de unos días en Estados Unidos. Antes sueno como George Lopez." (My grandmother had "perfect English," my mom has "perfect English." My English sounds better after a few days in the United States. Before that I sound like George Lopez.) Esteban chuckled.

Esteban's chuckle showed some warranted hesitancy to Javo's "perfect English" as contrasted with George Lopez. First, Javo's "perfect English" was fraught with language ideologies about how people should speak. These ideologies intersect with racializing discourses about the way Spanish-speaking Latinxs *should* speak. As Arlene Dávila shows, Latinxs themselves often argue over what is the "correct" way to speak Spanish.[14] When they see others speaking "accented" Spanish in the media, for example, they feel embarrassed; many feel a respectable Latinx-identifying person should be

14. Dávila, 2001.

able to speak either English or Spanish in its "pure" form in order to portray Latinxs in a "positive image."[15] Different Latinx groups call attention to the supposed faultiness of their language practices by demeaning "accented" speech and bilingualism of others, but "the irony is that all of these insidious distinctions fall short of challenging, and in fact reinscribe, the preeminence of whiteness and of the 'non-ethnic' as the abiding reference against which each of them is rendered suspect."[16]

Speaking "correctly" in this context thus means not only *not* code-switching between the two "correct" forms, but not speaking with an accent in either. Latinxs who grow up bilingual are often placed between these two, and bilingualism becomes equated with programs for "Limited English Proficiency" or "English Language Learners" in educational contexts.[17] Positioned alongside special education students as second-class educational figures, students are framed as proficient in neither English nor Spanish. They become "linguistically subhuman."[18] Jonathan Rosa calls this the ideology of "languagelessness": expected to speak two languages but understood to speak neither correctly, US Latinxs' linguistic practices are framed as "non-languages."[19] Mandatory language policies thus create stratified, class-based distinction between elite and remedial forms of bilingualism. That is, "while bilingualism is understood as a valuable asset or goal for middle-class and upper-class students, for working class and poor students it is framed as a disability that must be overcome."[20]

Thus, Javo's "perfect" English comment is driven by these ideologies, but the most revealing part of his remark was that he opposed his purported perfect English in relation to "sounding like George Lopez." George Lopez is a Chicano/Mexican-American comedian from Southern California who rose to fame in the 2000s with standup comedy geared toward a working-class Mexican-American audience. One of his recurring jokes is when he switches into a "white voice" by modulating the pitch of his voice: raising his pitch to index an "unhip" whiteness or lowering it to index a professional "voice of authority."[21] His stereotypical characterizations of whiteness, especially when indexed as a voice of authority, work to create a space where relations

15. Dávila, 2001: 195.
16. Dávila, 2001: 215.
17. Mendoza-Denton, 2008; Zentella, 2002.
18. Rosa and Flores, 2017: 624.
19. Rosa, 2016, citing Gal, 2006: 171.
20. Rosa, 2016: 171.
21. Fought, 2006.

of power are reversed, held up for inspection, and (most of the time) reified. His live performances function as a space where working-class Latinxs can "be themselves," complete with "accents" and all.

Coincidentally, George Lopez is from San Fernando, the same neighborhood in Los Angeles where Esteban was born and grew up. Perhaps there was some misrecognition on Javo's part, and he didn't realize that Esteban in fact sounded *like* George Lopez. Or perhaps his comment was geared directly toward Esteban, or more precisely, toward distancing himself from Esteban. That is, Esteban represents not only a version of English that isn't "perfect," but also part of the "backward culture" that still hasn't fully assimilated to the US ideals of efficiency, openness, and innovation, those represented fully by the Silicon Valley "culture" that Javo and his friends came to promote in Mexico in order to change the "Mexican chip."

Curious about these dynamics, I asked Javo in an interview how he felt about Latinx politics, framing my question around issues of diversity and access in Silicon Valley. Javo himself had participated in events aimed at increasing diversity and "empowering" the Latinx community on his trips to San Francisco. The event in the opening of this chapter is one example; he had also participated in summer-long bootcamp designed specifically for Latinx tech startups (qualified as both US Latinx startups and Latin American startups). Javo hesitated when I asked the question. "How do I say this. . . . I'm trying to be politically correct, I need to avoid hurting people's feelings," he tells me in his "perfect" English. "I totally disagree with the premise that you should have extra benefits because you're Latina or you're Black or you're a woman or you're gay or whatever. I totally disagree with that because I think it should be based on talent." Javo confirms his position in relation to minoritized populations in the US, which he lumps into "Latinas, blacks, gays, women, and 'whatever'." Here, very explicitly, Javo adopts the Silicon Valley concept of "finding talent," of colorblind meritocracy that finds those who are able to acquire the necessary skills and cultural capital to "succeed" and to blame people from underrepresented backgrounds for failing to do so.

Sound familiar? The logic that Javo uses parallels the same position he has toward all Mexicans. His comments are in line with the government's proposal to remodel Mexicans' chips and with his panel colleagues' wishes to build a "new" Mexico with new attitudes, new outlooks, and a new "culture." Later he confirms his vision for this "new" generation in Mexico:

Our parents' generation feels defeated, *son la generación de 'no se puede.'* We are the generation of '*si se puede.*' That generation was brought up around a sense of failure. Like, you're Mexican, you're not worth anything. We need to weed out that generation as soon as possible. We need to work on the small victories. We can actually achieve change if we change the way we think; if we stick together, we can see results and achieve what we want. This younger generation needs to forget about what we've always been told about Mexico, and we just need to create a new Mexico.

Javo's comments might appear to overlap with the "disenchanted generation" scholars have found in Mexico, in that he has appropriated neoliberal discourses about taking initiative, being self-satisfied, not waiting for government, and being "socially conscious."[22] But when paired with his comments about underrepresented communities in the US, in particular the Latinx "community" he selectively connects and disconnects himself from, he also aligns himself with hierarchies of global capitalism that encourage him to perform his country's modernization. This corresponding "coming of age" is always assumed in relation to other Latin American countries,[23] and in this case, in relation to other members of the "Latinx community" that might be holding him back. If Javo escapes the "specter of the Indian," the idea that the Indio part of the mestizo is the uncivilized, primitive, and incommensurable "Other" that permeates the "future-thinking" mestizo and prevents the "cosmopolitan" mestizo and the nation from becoming fully "modern,"[24] then here he employs the same logic to distance himself from the "Chicano," or the "Mexican-American" as the forward-thinking cosmopolitan Mexican.

Judith Irvine and Susan Gal call this "fractal recursivity," when binary oppositions are projected across scale onto another level of group structure or subcategory of said group.[25] Using this framework of fractal recursivity, Norma Mendoza-Denton further finds that young Latinas in California construct markers and corresponding ideas about "*norteñas*" (northeners) vs. "*sureñas*" (southerners), where the latter are seen as poor, unsophisticated newcomers.[26] The broader processes of racialization and corresponding language ideologies, as well as the capitalist structures that frame the Global North vs. the Global South get projected onto these "hemispheric

22. García Canclini and Cruces, 2012; Urteaga Castro Pozo, 2012.
23. Dávila, 2016.
24. Leal Martinez, 2016.
25. Irvine and Gal, 2001.
26. Mendoza-Denton, 2008.

localisms."[27] Thus, his moves across the techno-Borderlands, the positions that Javo takes to align himself with these subtle forms of difference-making, are in tune with his moves away from the Indio in his mestizo status, always framed in relation to an "Other" that also shifts, especially when crossing nationalized borders.[28]

Javo's statements and, quite frankly, his unpolished ways of thinking about language, class, and racialization across the US/México border unfortunately paralleled similar discourses and ways of creating social hierarchies that I found across my field sites. In a "Latina entrepreneurs" panel held in the Bay Area, for example, one young Latina-identified woman reprimanded the US Latinx community for not taking advantage of the entrepreneurial opportunities available to them. She herself was the founder of a startup consulting firm in the Bay Area. She highlighted her proven trajectory of "taking matters into her own hands" and not waiting for permission from anyone as crucial to her success. "We held an 'open office hours' for Latinx startups and guess who showed up?" she asked the audience. "Only people from Mexico, again," she responded to herself. She gave numerous examples of how the US Latinx community was not taking advantage of the "opportunities" available to them, and that *we* needed to step up. Except for the Latinx *makers* from the previous chapter, perhaps, the rest of us were failing to "catch up." We were refusing to become the modernizing tech entrepreneurs of the knowledge economy.

Thus, the intra-Latinx dynamics that have characterized working-class Latinx "frictions" in the US come to reconstruct themselves in the high-tech knowledge economy. As Nicholas De Genova and Ana Ramos-Zayas find, for example, working Latinxs enter into hierarchies of "deservingness."[29] These hierarchies are informed by politics of labor vis-à-vis the nation-state and result in unfortunate stereotypes of one another (e.g., Mexicans viewed Puerto Ricans as lazy and Puerto Ricans viewed Mexicans as submissive, "third world" people). These intra-Latinx distinctions perhaps become hyper-visible when it is time to organize around a specific social movement and make public demands, but they are "created in the everyday, in terms of ideologies of deservingness based on dignity, civility or modernity, gender and sexuality, or language, these disparate themes tended to be orchestrated through intersecting rubrics of racialization and inequalities of citizenship."[30]

27. Mendoza-Denton, 2008: 130.
28. Yeh, 2015.
29. De Genova and Ramos-Zayas, 2003.
30. De Genova and Ramos-Zayas, 2003: 16.

Within the code worlds I have been examining, citizenship also plays a factor, as those "Latinxs" who enjoy the privilege of citizenship are seen by some as failing to capitalize on opportunities and therefore not quite measuring up in terms of "talent" to those from outside of the US nation-state. As a transnational Latinidad is constructed (or prototyped) in these hacker and tech spaces, who *belongs* in the categorization of "talent" is up for grabs, as are the positions of the gatekeepers of this new privileged space.

I also don't mean to condemn Javo, or others who share his views and positions, for adopting these "neoliberalizing" discourses; my purpose is to understand the politics and infrastructures in which they are entangled. In the case of Javo, we learn that his family was adamant about teaching him "perfect" English, and that his great-grandmother was in fact from El Paso, Texas. As scholars of the Chicano Movement have shown, social movements are made up of different (and competing) ideologies and practices across regions. In Texas, for example, Arturo Rosales shows how activists took a more pragmatic approach, less ideological than California, because they were hungrier for material rather than cultural rights.[31] Chicanos in Texas were more willing to be part of a middle class where they could work "within the system" to promote social mobility amongst Chicanos, particularly because the line between Anglos and Mexicans was more visibly drawn. California activists, on the other hand, were confronted with more of an "identity crisis," where they were quick to reject middle-class lifestyles and experiment with alternative identities that pulled from African-American activist influences, a pre-Columbian past, and street youth culture more than folk mexicanismo.[32] While there might be some homogenizing of "California" and "Texas" communities on the part of Rosales and erasure of women within these spaces,[33] my point is to highlight how the shifting politics of race, class, and nation are more responsible for Javo's moves than his personal statements or subjectivities.

In other words, my aim is to show how racialization inflects theories of flexibility and the neoliberal knowledge economy, less than finding distinctions between Latinx collectives (or self-identified Latinx subjects). Here I share De Genova and Ramos-Zayas' more optimistic view for solidarity:

> Thus, if we have taken such great pains and gone to such extraordinary lengths to analyze the bases for Latino division, it has been motivated

---

31. Rosales, 1997.
32. Rosales, 1997: 224.
33. Blackwell, 2011; Ramírez, 2009.

by a more fundamental desire to explore the possibilities for effectively sustaining various ideas of Latino community and coalition that could viably serve to promote counterhegemonic sociopolitical projects formulated in terms of Latinismo.[34]

In the next section, then, I retake the concept of the pivot that I started this chapter with in order to discuss how the neoliberal logics underlying Javo's moves, these shifts and flexibilities across the techno-Borderlands, have the potential to work in favor of global capitalism.

[4] Flexible Neoliberalisms, Precarious Pivots

"Mexico Day" is just one instantiation of what the "tech startup boom" looked like in different parts of the world during the mid to late 2010s. My aim across the chapters of *Code Work* has been to highlight the heterogeneity of sites and subjects within Mexico itself, and across the US/México border, paying particular attention to dimensions of class, gender, and racial inequality, but also to point to the larger political economic processes that are not bounded by location or a particular nation-state or national border.[35] In other words, the challenge is to highlight what might be different about Mexican hacker-entrepreneurs at the same time that we pay attention to how they perform their belonging to a "global" hacker community. How do hacker-entrepreneurs navigate these national, racial, and ideological lines, as well as other dimensions of difference across the techno-Borderlands, as they attempt to construct and manage pockets of autonomy within and across the spaces and institutions in which they participate?[36]

Javo and his hacker-entrepreneurs might very well be answering the call to develop the "culture of risk" that modernizing, nation-building narratives so desperately call for. Their moves across the borderlands—cultural, national, and technological—might be precisely what is expected of subjects who work with the "new spirit of capitalism." As Luc Boltanski and Eve Chiapello outline in their book of the same title, this "new spirit of capitalism" fosters commitment and enthusiasm through management techniques that stress versatility, job flexibility, and the ability to learn and adapt to new

---

34. De Genova and Ramos-Zayas, 2003: 215.

35. Gupta and Ferguson (1992) spearhead the move away from producing studies that frame groups as bounded by "culture."

36. Coleman (2017b) proposes the rubric "weapons of the geek" to point to a shared set of cultural practices, sensibilities, and political tactics that connect diverse "hacker" communities.

duties.[37] In this sense, they not only work within the capitalist system but also help to construct it.

Hacker-entrepreneurs effectively contribute to capitalism-in-the-making; "capitalism" in this sense grows out of a particular set of institutional worldviews, subjectivities, and practices, and spreads under certain conditions at specific moments. The Wall Street investment bankers Karen Ho works with not only create markets, but immerse themselves in the market—their skills and social lives take on the anxiety, reinvention, and risk that characterizes it.[38] Similarly, disciplined commodity traders develop the sense that they they can "experience the market and become part of this living thing, intimately connected to it."[39] There are productive dimensions to "risk." Risk-taking becomes established and sustained by routinization and bureaucracy; it can become "objectified."[40] Risk can also become the celebrated skill of the "strong individual" who knows how to manage risk and calculate his current market value.[41] The code-switching entrepreneur who embodies risk thus performs flexible knowledge practices in order to gain "global technology expert" citizenship.[42]

Across the chapters of *Code Work* I've argued that my research participants use the underlying logics of software design to help them rethink social and political relations. A common critique of this argument is that these subjects are moving and acting *exactly* the way capitalism and neoliberalism expects them to move and act. But first we have to unpack "neoliberalism." Integrating ethnographic fieldwork into the computing stack, one of the aims of the "ethno-stack," functions as a methodological answer to decipher the global processes that constitute "capitalism" or "neoliberalism" on the ground. This approach moves away from the all-encompassing neoliberalism "package" and instead thinks about the neoliberal as a logic of governing for optimal outcomes, an array of techniques that is mobile, abstractable, and flexible as it migrates from site to site, interacting with various assemblages that cannot be analytically reduced to "neoliberalism."[43]

Moreover, whether hacker-entrepreneurs can be categorized as "neoliberal subjects" might be largely irrelevant. Anthropologists have shown how

37. Boltanski and Chiapello, 2007.
38. Ho, 2009.
39. Zaloom, 2004: 379.
40. LiPuma and Lee, 2004.
41. Miyazaki, 2006.
42. Ong, 2006.
43. Collier, 2009; Ong, 2006.

neoliberal logics can be used in ways that contradict the negative connotations usually associated with neoliberalism. Lisa Hoffman, for example, shows how young professionals in China replaced bureaucratic job assignments with labor markets to produce a self-enterprising ethos, even as they acted in the name of a patriotism representative of the Maoist era.[44] In Guatemala, Monica DeHart finds that indigenous activists invoked norms of efficiency, transparency, and accountability—all associated with neoliberalism—precisely to criticize state policies frequently characterized as "neoliberal."[45] Likewise, Javo first proposed infrastructure for combating voting poll fraud by organizing an efficient, transparent, accountable group of activists to intervene in state practices he and his colleagues deemed corrupt.

A more productive approach than pinpointing what is "neoliberal" or not, then, is to understand how these logics can sometimes provide subjects the tools to think about their particular situations. In her research with job-seekers in the tech industry in the Bay Area, Ilana Gershon explores the moments when US Americans face contradictions when implementing these neoliberal logics; she focuses on the moments when workers must be open to finding different ways of being social beings.[46] In an industry where striving to be a full-stack developer means constantly keeping up with the latest technological infrastructures and learning to sell your skills at each level of the stack to different employers, finding a job as a software developer in the tech world is not easy. A worker must be flexible, but not too flexible; they must take risks, but not too many. In addition to these entrepreneurial conundrums, hacker-entrepreneurs from Mexico must also consider how hacking and code work are valued for some, but undervalued or even criminalized for others. Hacking from "the South," combined with entrepreneur-ing from the South, adds a plethora of even trickier situations than those working within the confines of one nation-state must face. This precarious terrain calls for even more careful pivots.

"Flexibility," then, and closely aligned pivots in the techno-entrepreneurial world might be framed as ideal, but only within very particular limits, as code workers manage commitments and networking with a perpetual renewal of "skills" that they must bundle into legible packages which help to define their worker, coder, and hacker selves. They must learn to couple or un-couple these selves and politics across the computing stack, or as I've argued, with the personal, interpersonal, and especially with the sociopolitical layers

44. Hoffman, 2010.
45. DeHart, 2010.
46. Gershon, 2018.

of the ethno-stack. As workers manage this complicated game, "Moment after moment, people will continue to wrestle with these dilemmas anew, choosing to go in one direction in one instance and in another, perhaps contradictory direction, in the next instance."[47] The lived dilemmas of being a "neoliberal self" become the reasons why people begin to reject neoliberal logics or transform them into something else entirely, Gershon argues. Of course, as is the case with most of these research studies, scholars tend to focus on abstract, generalized workers who lack any markers of difference. As I've attempted to tease out throughout this chapter, theories of the neoliberal and corresponding flexibility become more complicated when we take into account nationalized and racialized borders.

[5] The Latinx Hacker-Entrepreneur Pivot

We met Javo at the opening of the chapter firmly presenting himself and his startup as Mexican at the "Mexico Day" event in Silicon Valley. As I've demonstrated throughout this chapter, though, Javo learned to become "Mexican" here but not there, "Latinx" there but not here, as he accepts the title of a hungry Mexican/Latinx "talent" but disassociates from the figure of the marginalized Latinx when this association works against him. As Arlene Dávila argues, the very fluidity of a constructed Latinx identity becomes a commodity.[48] Dávila's analysis is grounded in a simple observation: how could it be that Latinxs are celebrated in contemporary media and public representations for their "culture," for their "coming of age" in America, while at the same time they are represented as an economic liability who take jobs, resources, and benefits from "regular" Americans?

Less interested in whether these representations are accurate or not, Dávila examines this Latinx "spin," the selective publication, circulation, and deployment of Latinidad, how and why some representations come to dominate over others, especially more marketable ones. Dávila reveals the nuances of this "Latinx spin": Latinxs are sometimes represented as "giving America back to America," as orderly, sanitized, respectable middle-class, employed, family-loving citizens, when in fact many Latinxs find themselves forming part of working-class enclaves, lagging in education, wealth, and access to services and infrastructure.[49]

47. Gershon, 2018: 175.
48. Dávila, 2008.
49. Dávila, 2008: 8.

If these selective representations and circulations of Latinidad point to the Latinx "spin," then Javo's moves across the techno-Borderlands index the Latinx "pivot." That is, the pivot, a term used to guide entrepreneurs to move with the market, to follow the trends and construct a product with an audience and corresponding consumer-base, is now a tool to think with (and move with) as the investors who participate in the labor and knowledge economy spin their own versions of "talent," which they map onto shifting racializations of collectives. As the examples I have presented in this chapter show, Javo plays with the visibility/invisibility of his startup, wears the sombrero here but not there, and avoids speaking with a particular English accent in an attempt to dodge the racialized politics that prevent him from advancing his startups, whether these projects revolve around politics or around pizzas. As he traverses these structures, the pivot becomes another tool to think with, about the way the knowledge economy is structured, about the way "communities" fit into them, and about how one might be able to, at least for the moment, construct (or "hack") a pocket of autonomy within these processes and infrastructures.

Ethnographic work in the code worlds means diving into the hacking but also shuttling out to other dimensions of life that are sometimes removed from the code worlds, and many times intimately connected to them. Javo and his co-founders identify as hackers and are active participants in hackathons; they're also mobile entrepreneurs who transport their code work across nationalized borders as they meet investors and other tech world representatives beyond the code worlds. As they learn to navigate shifting relations of power, they become enmeshed in multiple and contradictory language ideologies and racializing processes. They also learn how to project these same ideologies onto others as they attempt to use concepts such as the pivot to make moves that presumably counter the flexibility and the pivots performed by the companies and the investors themselves. While they form part of the collectives that aim to hack processes that portray them as Othered hackers, they're also caught in practices that work to erase *other* differences. The geographic naturalization of "perfect English," for example, leaves no room for multilingualism, Spanish-dominant bilingualism—ultimately, no room for the presumed "unmotivated" or "non-entrepreneurial" Chicanxs or Mexican-Americans. They're caught replicating the same discourses and structures that construct deficient Latinx subjects who are purportedly stuck in a phase where they're always trying to catch up, forever searching for a way to stay ahead of a game where the rules are ever-changing and never in their favor. Whether their code work is rooted in social justice or revenue generation—that is, whether they're hacking difference, hacking corruption, or hacking pizza deliveries—the racialized, classed, and gendered borders are always present.

An update on Javo's professional trajectory and his pivots ends this chapter on a hopeful note. In the techno-entrepreneurial tech startup world, a commonly held belief is that there are two proven paths to success: either you're a visionary who can see needs and solutions nobody else can, or you're a steadfast entrepreneur who pivots their way to success. Javo certainly went the latter route, and adopted other elements of the Silicon Valley narrative as well. He tells the story of how he and his other two co-founders at the time had to work cleaning the hacker hostel where they were given the opportunity to stay in San Francisco. They stretched their $150 USD/month grocery budget (for all three) to afford to cook one meal a day, while completing their subsistence by attending as many free tech events as possible in order to load up on as much free pizza as humanly possible; a classic "pull yourself up by our bootstraps" Silicon Valley narrative, reminiscent of tech entrepreneurs who hack their nascent ideas from a garage with very few resources. When asked about his various pivots, he also says things like, "We kind of had to pivot. We didn't choose to pivot—users were telling us this is what they needed," and repeatedly asserts that he and his co-founders have always been focused on building the best product possible, reinforcing this with statements such as "We don't support or not support anyone; we are just a tool." These statements align with the way Silicon Valley seems to always find ways to try to de-couple politics from their technologies, but Javo has had to voice these responses precisely because his pivots eventually did lead him right back *to* politics.

One of the final major pivots of his app was Pingafy, which used Bluetooth-enabled mesh networks to send messages between users without needing an internet connection or cellular network. The idea for this pivot, Javo tells various audiences, came when he and his co-founders were headed to the US on the startup bus (the same startup bus from the introduction of *Code Work*), and they experienced intermittent connection on the bus, in addition to having just lived through several earthquakes in Mexico City that left residents disconnected and unable to communicate with one another. They ultimately won the startup bus competition, but after much effort attempting to get this "connecting the unconnected" app funded, Javo and his team realized that perhaps they had something much more powerful in their hands than just an app. Instead of selling their app as a unit, on the higher levels of the computing stack, they could sell the underlying protocols of the mesh networking that made the wireless communication between mobile phones work. After months of trying to get it to work, and even reaching out to developers of major US tech companies and receiving no responses from them, they had hacked the networking protocols to get the wireless messaging system to a functional state. This design, they realized, could be packaged into an SDK (software developer kit) and sold to other app developers who wanted to add this particular functionality to their own apps. They advertised it as a novel way to "gain independence from third-party service providers." The independence they had gained by learning to infiltrate the computing stack could now be packaged and offered to others. In this sense, it could even be used to empower communities of software developers from the Global South.

But their strictly "product-focused" goals got away from them, as the app quickly escaped into the hands of users they had not foreseen, in surprising and arguably much more impactful ways. Pingafy experienced sharp usage spikes in not only delicate situations around the world, such as earthquakes and hurricanes, but also in areas of social conflict, where protesters in countries such as Myanmar, Hong Kong, and Ukraine during the late 2010s used it to communicate with each other and organize their efforts away from government interference. Javo and colleagues took notice and quickly presented their underlying technologies as the solution for any other app that wanted to implement communication "outside of the public eye," a feature that was becoming increasingly popular even with major mainstream communication apps. This led to Pingafy attracting the attention of angel investors, who ultimately invested millions in Javo's company, allowing him and his co-founders to hire more engineers and continue to work on what they then referred to as a platform (instead of just an app), and what wide-eyed investors were referring to as a "disruptive technology." Javo's statement to me, many years back at Mexico Day in Sunnyvale, "Para tener impacto, necesitas feria," (To make an impact, you need money/bling,) had apparently come full circle.

Focusing on the root layer of the computing stack, close to those underlying logics and protocols that drive the code worlds, seems to have helped Javo finally have his code work valued by investors and to have his Pingafy app ultimately have an impact in communities and politics that he was originally interested in, but that had little traction in the world of techno-entrepreneurial Silicon Valley. To be clear, his original anti-corruption app could be described as a "technological fix" in line with a type of neoliberal politics that works precisely to de-politicize politics. But even if it wasn't the most sophisticated way to resolve voting fraud, his sight was originally set on protesting the status quo, on raising his and his colleague's voices, on aiming their code work at social injustices they wanted to remediate. Their app might have been initially pivoted away from politics to pizzas when they encountered entrepreneurial Silicon Valley practices, but it *was* eventually used for collective organizing and protesting. How did this happen if they assimilated the discourse of de-politicizing tech, of focusing only on your product and pivoting with the market?

By keeping their head down and working on the product, on the technology, on the root layers of the computing stack, Javo and friends avoided the dilemmas that have plagued the cast of code workers I've presented across this book. They didn't have to decide whether they should be "successful" or whether they should work on empowering their communities. They didn't have to deliberate whether their politics had to be tightly coupled or loosely coupled across layers of the ethno-stack. And they didn't have to worry about whether they were compromising any of the particular hacker ethics that defined the hacker portion of their hacker-entrepreneur identities. But ultimately, their final surprising pivot, and the "politics" to which its app eventually returned, are inextricable from the experiences they accumulate as they think with the code and with techno-entrepreneurial language—the pivot—to navigate boundaries of nation, race, ethnicity, and class across the techno-Borderlands. These differences are resurgent at every step and call for creative traversals that combine the market logics of agility and risk with the hacker logics of reinvention and resistance. Their responses to tricky neoliberal predicaments are heterogeneous, "flexible," and strategic—not conditioned, unthinking responses or simple acquiescence to "the system." Pivoting identities, presence, and language points to migration itself as a type of hack that hacker-entrepreneurs learn to embody. In my coda, I continue to explore this notion of migration and movement across borders as a type of hack to close the loop from the book's introduction on how to cultivate border-code-workers.

[7] Coda: Working Code AND Working Futures

Working the code inherently involves thinking about imposed borders in order to rethink and reconstitute boundaries and the interfaces between them. Similarly, across *Code Work* I've also attempted to transgress disciplinary borders by connecting perspectives, theories, and interventions from across academic disciplines that are sometimes siloed from each other. In an effort to connect my research findings and methodological proposals to audiences who might approach migration, labor, technology, and transnational Latinidad from different perspectives, I end by returning to a film that has increasingly captured the imagination of scholars across academic disciplines.

In the 2008 Mexican science fiction film *Sleep Dealer*[1], we meet Mexican migrant workers who work in the US. Their labor is recognizable as "unskilled" migrant work; they perform manual, arduous jobs. One man works for hours nonstop in the fields while another looks down for a moment from a high-rise building as he helps place the steel beams necessary to finish the construction project. What separates them from contemporary (and past) migrant workers is that these workers of the future have not left Mexico—they are the "node workers" who use implants, or "nodes," to control robots, and these robots perform the manual tasks for them. The film's concept was first introduced by the director, Alex Rivera, in a short satirical piece *Why Cybraceros?*[2] The video circulated on the internet in the 1990s and used actual footage from a propagandist 1959 video *Why Braceros?* which promoted the mid-century Bracero Program. The program can be described as a series of laws and diplomatic agreements which allowed temporary workers from Mexico to provide manual labor in the US.

1. Rivera, 2008.
2. Rivera, 1997.

In the video we hear the official pitch for the program: "As agriculture has become a larger and larger industry in America, it has become harder and harder to find American workers willing to do the most basic farm tasks. Picking, pruning, cutting, and handling farm produce are all simple but delicate tasks." Rivera's short video plays off of this pitch and connects it with 1990s internet utopianism to propose a new version of the bracero, the *cybracero:*

> In Spanish, cybracero means a worker who operates a computer with his arms and hands. But in American lingo, cybracero means a worker who poses no threat of becoming a citizen, and that means quality products at low financial and social costs to you, the American consumer.

The cybracero thus resolves the "problem" for both Mexico and the United States. For Mexican migrants who aren't granted citizenship, they no longer have to suffer from the forced covertness they are subjected to, or from deportability, the sociopolitical condition in which physical removal from the US nation-state is a constant worry.[3] From the US perspective, the scenario is captured succinctly by one of the *infomaquila* workers in *Sleep Dealer*: "Le damos a Estados Unidos lo que siempre han querido: todo el trabajo, sin los trabajadores" (We give the United States what they've always wanted: all the work, without the workers).

Academics across disciplines have honed in on the themes of migration and technology that the film raises. Curtis Marez places the film in a category of speculative practices and "farm worker futurism" that grows out of Californian agricultural regions, especially when agribusiness corporations and farm workers (and their unions) debate the transformations that new technologies bring to labor and production.[4] B.V. Olguín takes issue with the film's individualist takes on true systemic, revolutionary change: "It leaves no room for harnessing technology as part of a coordinated revolutionary struggle."[5] This individualist positioning in relation to the liberatory potential of technology is best represented by the main character Memo's final quote, where he has accepted his "becoming one" with technology: "Un futuro con

---

3. Leo Chavez (1992) focuses on the condition of covertness in undocumented migrant communities as a response to the risk of being deported. Nicholas De Genova (2005) shows how migrant "illegality" signals a specifically spatialized sociopolitical condition. He proposes that "illegality" is lived through a palpable sense of deportability, the possibility of being removed from the space of the US nation-state. Victor Talavera et al. (2010) expand on this notion of deportability to show how it is omnipresent in migrants' everyday lives, discursively, materially, and experientially.

4. Marez, 2016.

5. Olguín, 2017: 136.

un pasado, si me conecto y lucho" (A future with a past, if I connect and struggle). To inhabit the promised better future, he must connect himself using the nodes and struggle *with/using* the new technology.

The film has recently gained prominence in academic studies as well as academic classrooms because it provides much material for analyses of the politics of labor between the US and Mexico, as well as the relation between humans and technology more broadly. As one node worker connects within the infomaquila in Tijuana, "city of the future," he states, "A veces tú controlas a la máquina, y a veces la maquina te controla a tí" (Sometimes you control the machine, and sometimes the machine controls you). Raising issues of technological obsession and human alienation, Luz tells Memo of her ex-boyfriend: "La tecnología resultó más interesante que él" (Technology turned out to be more interesting than him).

While the movie contains endless one-liners and subject matter that can be unpacked for academic discussion, one very brief image—most likely ignored by most viewers—immediately jumped out for this anthropologist focused on "hacking" as a critical site of inquiry. During the opening of the film, we find Memo in a small shack tinkering with circuits, speakers, and unspecified electronic gadgets. For a brief moment the camera focuses on a voltage measuring device, which is placed on top of a book titled, *Hackear para principiantes* (Hacking for beginners). We learn that Memo has been hacking into drone communications transmitted by a multinational corporation that has effectively eliminated water from the area where Memo lives in Santa Ana del Río, a small town in Oaxaca. Memo describes Santa Ana as "una trampa: seca, sola, desconectada" (a trap: dry, alone, disconnected). He's interrupted by his father, who takes him out to work the *milpa* (maize field), and who seems to be annoyed that Memo spends too much time with his tinkering and not enough time outdoors. "Por lo menos sé que el mundo es más grande que esta milpa" (At least I know that the world is bigger than this *milpa*), Memo tells his father.

This brief hacking scene is important because it confirms many of the imaginaries the general public has about hacking. There's a man in front of a computer (or objects that resemble technological gadgets) infiltrating some system, either to extract secret information or to carry out some sort of malicious deed. The infiltrator is usually some highly talented (or highly awkward) young man who can do magic with his technical abilities; usually, he accomplishes this from his parent's basement. If the image doesn't fit the stereotype of this (usually) white and (usually) middle-class (almost always) male, then it's usually an Othered hacker from an exotic location whose ability is threatening to the integrity of some system.[6] Anthropologists who have taken hacking as their anthropological object, after following the object across multiple borders, and consolidating other studies on hacking, announce:

> Hacking exists: whether it is referred to as leaking or breaching; whether it involves state actors, criminals or anarchist activists; whether it seems to disrupt an election, protest a corporation or government, or steal funds; whether it is about making software in a different way, or breaking it in a new way, hacking is a here to stay, whether we want it or not, and we learn more about it, the more carefully we look at and study it.[7]

Interdisciplinary scholars have followed state actors, hacktivists, FLOSS (Free/Libre and Open Source Software) developers, hack-driven leakers and journalists, criminal extorters of bitcoins, information security researchers in search of a safer internet, and in my case, what I call "hacker-entrepreneurs," in search of who the hacker is, what the hack is, and what exactly it means to hack. Across my chapters I've explored how Mexican and Latinx hacker-entrepreneurs position themselves in relation to projects *in the name* of hacking, and especially projects that frame hacking in terms of community empowerment, whether this community is a particular racialized or gendered group, or a nation. We've met a cast of code workers who construct strong senses of self while they negotiate their sociopolitical realities, all while honing their coding practices, interacting with and "thinking with" the underlying technological infrastructures that make up the code worlds. In *Code Work* I've mobilized ethnographic research to determine what hacking means to people, how they practice it, and why it's important for the global information technology economy broadly and for US/México transnationalism specifically.

While the researchers who have been fascinated with *Sleep Dealer* focus on the techno-dystopian aspects of the film, especially the way that technology transforms labor and migration, an under-examined aspect of Rivera's work is precisely this transnational component of the film's plot. That is, the two main characters, Memo and Rudy, are two very differentially positioned subjects who end up coming together and combining their respective abilities and stories to counter a large, multinational, militarized corporation that

---

6. Especially with the events following the US 2016 Presidential Election, when the US Department of Homeland Security reported that Russian "hackers" had meddled in process, the "Russian hacker" is particularly legible.

7. Coleman and Kelty, 2017.

does not necessarily have the livelihoods of farmers like Memo and his family in mind. Rudy is Chicano/Mexican-American and a newly minted drone pilot for this corporation;[8] Memo is a small town hacker whose father was killed by the same corporation. The twist is that Rudy is actually the person responsible for killing Memo's father in the drone strike, and he commissions Luz, the third main character of the film, to find out more about the man he killed. Rudy ultimately finds Memo, and after some intense moments and confrontations, they slowly develop solidarity. Rudy then uses and gets rid of his cyborg extensions to destroy the dam he once protected in Memo's hometown. Water returns to Santa Ana and the multinational corporation receives a devastating blow.

It is clear that Memo decides he needs to struggle *with* the machine in order to fight the system, and Rudy develops this consciousness about his wrong doings through technology also: Luz uses nodes to upload her memories to the "True Node" website, which Rudy uses to learn about Memo and the innocent people he is killing. Thus, there's a degree of political consciousness that arises in both of these hackers in order for them to accomplish their goals, and this consciousness is born out of a transnational collaboration.[9] Memo is from a small town in Oaxaca in Mexico, a location more easily recognized as the "Global South"; Rudy is a Mexican-American/Chicano, a racialized population in the US.[10] The coming together of these experiences "from below" are fundamental in the struggle against oppression in the search for alternatives, especially when they occur within or across the Global South.[11]

I end with these scenes from *Sleep Dealer* and my corresponding commentary because they're useful to analyze why I've decided to focus on "hacking" in *Code Work* and why it's important in contemporary life. Hacking has become a widespread cultural phenomenon because it provides humans with the practices, narratives, and imaginaries to think about their relationship with technology, and how this relationship might be leveraged to work "against the system." Within the hackathons, the hackerspace, and

8. While Rudy might not necessarily identify as a "Chicano" during the film, he's positioned (by the director and most likely by audiences) as a Chicano in that he is born and raised in the US and speaks a form of Spanglish when he meets Memo.

9. Indeed, Rudy is also positioned as the stereotypical hacker, if only by the scenes where he appears as a hooded, mysterious figure ready to use technology to accomplish covert operations.

10. Perez (2015) shows that Latinx and other youth of color have increasingly joined the military in an attempt to be recognized as full citizens and to benefit from the positive associations and respect that the military garners in American public life.

11. Santos and Meneses, 2020.

the co-working spaces, we find the hackers, hacker-entrepreneurs, and also "normal," everyday citizens (who shuttle between and blur these identities and positionalities) who plug into and disconnect from the joys and possibilities of hacking for myriad reasons. Popular discourse increasingly tells audiences that "anybody can hack," whether it's the Silicon Valley narrative that tells young middle-class kids they can hack from their garage or their parent's basement in order to start a company and become a tech multi-billionaire, or whether that "anybody" is someone like Memo, who can hack from their small shack in their village in order to fight the system. As I've argued in *Code Work,* the "promise of technology" inherent in hacking is one many of my research participants readily circulated: from El Chico Partículas' comment that perhaps someone from *la sierra* (the mountains of Mexico) could be the next computer genius, to the organizer of the all-women's hackathon who proposed that hacking could empower even those who are entangled within intersectional vectors of difference, to the Javos of the tech startup world who pivot their identities, mobilities, and apps until they get their technologies to do the political work they envision.

Returning to the missed connections between the three collectives aboard buses that I unpacked in *Code Work*'s introduction, I've teased out the moments across hackerspaces and hacker lives where such potentials for collaboration across difference actually crystallized, where they could have been further developed, or where they brushed up against each other in productive ways. In Mexico City, the loose definitions around "hacking" allow Leo and El Pato, who lie on opposite poles of the hacker-entrepreneur spectrum, to come together across classed boundaries to practice distinct versions of hacking. Despite their differences, they nonetheless immerse themselves in the code worlds and use coding concepts and metaphors such as "batches," "exceptions," and "loose coupling" to think with the code, about their relationships with employers, and with social and government institutions. At the women's hackathon, Mariana, Alicia, T.C., and the abuelitas come together to effectively rethink participation infrastructures, challenge strict demarcations between the technical/non-technical, the traditional/modern, and build intergenerational solidarity by inviting programming newcomers to align themselves with more welcoming layers of the computing stack. Migrahackers Cesar, Cindy, and Fernando seem to have been brought together by destiny, collaboratively hacking immigration politics they have been passionate about their entire lives, and mobilizing their distinct professional affordances to cultivate the hacker ethos, while they simultaneously transform the practices of prototyping and iterative code design to interrogate developing versions of a transnational Latinidad. Esteban, a self-identifying Latinx from the US, and Cofi, from Mexico, share experiences at the CDMX hackathon that reveal the contradictions of valuing the practice of coding as labor across constructions of class and masculinity, but they ultimately recognize the fascinating, complex intersections between the hacker ethic and the transnational migrant ethics of "hard work." But Esteban's own moves across the border find delicate frictions with the moves of Javo, who repurposes the tech startup logic of the "pivot" to pivot his identity, language, and presence as he selectively accepts the title of a hungry Mexican/Latinx "talent" but disassociates from the figure of the marginalized Latinx when this association works against him.

In these very different but connected ways, Mexican and Latinx hacker-entrepreneurs from all walks of life end up wrestling with what I call the techno-Borderlands. They experience what it means to navigate the political economy of tech and code work as it relates to US–México politics, and encounter the subtle subjectivities of difference found apart from yet deeply influenced by this border between the so-called "first world" and the so-called "third world." US/Mexican hackers thus deploy the language of coding to traverse intersecting boundaries of nation, race, ethnicity, class, and gender that remain constitutive to the transnational economy of tech and thus are reconstituted and refreshed at every step.

Clearly some code workers benefit more than others from the privileged mobilities that allow them to make these moves across the US/México border. Some simply can't find the resources that would allow them to travel to the US. Others such as Leo save up money all year from their coding gigs to make the annual pilgrimage to the tech company coder events. Others, even if they could save up the money, can't surpass (or refuse to partake in) the unwieldy process of receiving a US visa. And the lucky few, whether they were born in the US by result of migration from Mexico, or they possess the cultural capital to have obtained a US passport from Mexico, or as a result of a number of other life situations (too many to enumerate here), can travel back and forth to develop a privileged perspective on what things look like from both sides of the border. Of course, then, moving transnationally "requires both economic resources (money) and the legal-juridical capacity (passport, residency, citizenship) to move freely, which only some have. At the community level, however, the fact that the phenomenon exists and is enacted by more highly resourced community members has effect and implications for the community overall."[12] Thus, only some may be the physical travelers across the techno-borderlands, but their experiences affect the community of Latinx and Mexican hacker-entrepreneurs as a whole; their trials and tribulations iteratively reconstruct the techno-Borderlands.

Indeed, the cast of code workers I've presented are less like the international technological elites who strive to make connections with US elites despite any strained relationships between their countries,[13] and also unlike the "neither here nor there" migrants who escape the reach of the nation-state but are often framed as never seeming to interact with any space in any meaningful sense.[14] Instead, their transnational moves might be better described as aligned with a form of "diasporic dialectics," constantly in negotiation with institutions of citizenship on both sides of the border,[15] but also becoming versed and trained in binational politics that will guide them to an intimate understanding of the shifting, multifaceted techno-borders and their constitutive classisms, nationalisms, and racisms: that one might be racialized here but not there; that identities and credentials might be legible there but not here; that a particular political stance might not work in either and may need to be pivoted, for the moment.

Turning systems on themselves is what calls many of us to hacking. But what calls us to border hacking? Like the Mexican node worker in *Sleep Dealer* pondered, "A veces tú controlas a la máquina, y a veces la maquina te controla a tí" (Sometimes you control the machine, and sometimes the machine controls you). Human/machine. Worker/citizen. US/México. Border hacking offers the possibility to observe the games being played on both sides of these tenuously constructed yet fiercely imposed borders.[16]

For US/México hacker-entrepreneurs, travels across the techno-Borderlands begin to unveil what this "border hacking" or what "migration as a hack" might mean, always with the help of the code work, which occurs across scales and across layers of the ethno-stack. Thinking across scales, between layers of abstraction, identifies a skillful computer programmer, but

12. Candelario, 2017: 238.

13. As described by Yost (2017: 251), for example, drawing from Ross Bassett's 2016 *The Technological Indian*.

14. De Genova (2005: 98) succinctly dismisses both assimilation/incorporation and postmodern space theorists: "In one case, in the promised land at last, they might as well get down on their knees to kiss the ground; in the other, in a virtual world of their own, their feet never seem to touch the ground".

15. Félix, 2019: 8.

16. Beltrán, 2021.

it also helps to develop the mastery in finding those spaces or those moments where things come together—where politics, the social, and technology become entangled and their precise interconnections can be held up for inspection. This is reminiscent of the "border looping" that performance artist David Morison Portillo partakes in, traversing the San Ysidro port of entry at the US/México border repeatedly without stopping on a given evening.[17] "Es a través de la microdinámica de la interacción que pirinolear interrumpe el sistema" (It's through the microdynamics of the interaction that looping throws a wrench into the larger system), Portillo says about his performance protest.[18] In this exaggerated, accelerated version of "migration," he attempts to slow things down, expose, rarefy, find anomalies in and subvert the system, throwing himself at this exceptional space of the port, where the state has exacerbated power. About how his own ritual engages with the state's rituals, Rihan Yeh, his collaborator, and he say:

> David found presence necessary to lodge his complaint, make his disagreement felt—but in the port, as David said, there's no loitering. Looping began as a way to hack this prohibition; it was the only way David found to insist, immediately and in person, one, that there be soap and, two, that officers behave themselves. "Te vuelves imposible de ignorar" (You become impossible to ignore), he told me. Looping thus reveals how the port prohibits not just loitering but presence, el derecho corporal de permanecer ahí (a body's right to remain there) except in the most ephemeral way.[19]

Interestingly enough, Yeh and Portillo use similar language to US/México hackers: "looping," "hacking," and in this case using "presence" to make a simple demand that the soap dispenser in the bathroom be refilled, a request repeatedly ignored during his multiple crossings. Likewise, hacker-entrepreneurs in my final chapter find a way to hack the political economy of tech and its underlying machine of north/south hierarchies by pivoting their presence, by using their privilege to move across the border repeatedly to "loop" through apps, demo days, accelerator programs, and layers of the stack until something clicks, until the anomalies are discovered and things finally move in their favor, or at least until the injustices are exposed. They learn to pay attention to *differences* across the techno-Borderlands, and, as

17. Portillo and Yeh, 2021.
18. Portillo and Yeh, 2021: 35. Translations by authors.
19. Portillo and Yeh, 2021: 19. Translations by authors.

I've argued, learn to consider how politics, positions, and position-taking emerge from within spaces that claim to be de-politicized (by becoming a hacker or a maker) but also from spaces which are explicitly political (by becoming a Migrahacker). Across the techno-Borderlands, hackers and entrepreneurs develop new forms and practices of hacking that incorporate the market/neoliberal logics of competitiveness, agility, and risk with the logics they use in the code itself.

To become border-code-workers of the future, then, they must learn to connect the code work to the border work. The code work guides the metaphors, logics, and ethics that these othered hackers deploy across domains of social and political life. Across these diverse spaces and experiences, US/México hacker-entrepreneurs use their code work to develop heuristics for analyzing the organization of entities and relationships among them, whether they are elements in a coding environment, actors in a political-economic environment, or acquaintances in their intimate social environments. While I argue that this code work happens at the sociotechnical layer of what I've offered as the ethno-stack, I propose that connecting this work across its other layers—the personal, interpersonal, sociopolitical layers of the ethno-stack—will guide us toward thinking more holistically about the computing stack.

The ethnographic part of the ethno-stack prompts us to use the qualitative data gained from participant-observation and interviews, mobilize modes of reflexivity, and loop findings back into the ethno-stack. The ethno-component of the ethno-stack challenges us to develop analytics that privilege the construction of difference and politics of code work alongside the making and use of technology. And returning to the call of the Borderlands, border thinking, fundamental to the "border work," provides us a framework to "make connections among seemingly disparate events, persons, experiences, and realities," as well as to build on "holistic activist-inflected epistemology designed to effect change on multiple levels."[20] The border-code-work further allows for Anzaldúa-inspired scholars looking for an *in,* for doorways or for *bridges*, into the code work.[21]

Connecting the code work to the border work is important especially as techno-social initiatives increasingly claim to be rooted in promoting economic, racial, and gender justice. In Mexico, during a return trip to Xalapa in 2021, I found that iLab had been transformed from co-working/hackerspace

20. Keating, 2005: 8.
21. See Rosa's (2022: 437) proposal for "code ethnography," which summons Anzaldúa.

to political billboard. The space was plastered with political propaganda from the current party vying for representation in state elections. Then, just a year later, in 2022, it had transformed into "Educa," Mexico's "First Intelligent University," focused on innovation to create the professionals of tomorrow. This transformation of the iLab, now in private hands instead of the state's hands, seemed like a normal enough iteration for the hackers in Mexico who had become accustomed to the ever-shifting array of the state- and corporate-sponsored hackathons, hackerspaces, and co-working spaces that had endured from Peña Nieto to AMLO, as "men of the system" and "men of the president" reorganized themselves around new projects and so-called reforms. Perhaps this was just the latest manifestation of the "todos con el mismo chip" government initiative. Thinking with the state and with the market across their own iterations means carefully considering what is at stake with these models that center code, and code as a model for thinking about just futures, from the iLab to the intelligent university to the eternal reference point of Silicon Valley. The answer, which any burgeoning *border-code-worker* will know, is that the code work can never stand on its own.

And of course, my questions and answers always bring us back to the ethno-stack. How might *the stack* guide us toward understanding how structures of innovation always seem to re-assemble themselves into structures of inequality? How might the *ethno-stack* lead us to the necessary border work for participating in, observing, and dismantling these re-assemblies, within the hacker worlds and beyond them?

As this book demonstrates, code workers are trained to deploy a type of "systems thinking" that is always trying to reach the next layer of abstraction, always pushing the code work to move to broader domains of meaning. Social life works similarly. I propose that thinking with the ethno-stack can enable ethnographers, no less than code workers, to ground the stack logic of a variety of sociotechnical systems. While each implementation of the ethno-stack will call for different research questions across its layers, the following questions can guide our thinking with the generic ethno-stack:

How and why do people choose to enter the stack?

What are the different layers that make up this particular stack?

How do people use metaphors from one layer of the stack to solve problems in another layer of the stack, or another domain of life?

What type of work does it take to move between the layers of the stack?

How does difference in social position (along dimensions of race, gender, class, disability, sexuality) influence the way users navigate the stack?

How do other(ed) communities propose we reconfigure the stack?

What might alternative stacks look like?

If we understand technology as "always implicated in and shaped by social struggles,"[22] and as "always already social and always already connected to other technologies,"[23] existing in the broader domain of technology, then wrestling with the stack, teaching code workers to ask particular questions of and through it, and then integrating this knowledge formulated back into the stack is an entryway to effect change in broader social struggles. Whereas people often think of computational data structures as abstract, generalizable, and impermeable to difference, the ethno-stack asks how we can incorporate other(ed) concepts and structures as we think with the code, but also beyond it, and against it when necessary.

22. Sterne, 2003: 383.
23. Sterne, 2003: 383.

Glossary

App

Abbreviation for "application." In the context of the *hackathon,* refers to a software application, usually a small, specialized program for mobile devices (smartphones or tablets). An app can be made available for *users* to download and install through various internet app stores.

B/borderlands

In her genre-breaking writing, Gloria Anzaldúa develops the concept of the "B/borderlands." For Anzaldúa, the lower case "b" borderlands refers to the geographic region separated by the geopolitical Texas/Mexico border on which she grew up, and the upper case "B" Borderlands encompasses the psychic, sexual, and spiritual Borderlands of her own embodied subjectivities, resulting from oppressions experienced due to her culture, color, health, gender, sexuality, economic status, and especially her complex relationship to language. The B/borderlands, then, "in both its geographical and metaphoric meanings—represent intensely painful yet also potentially transformational spaces where opposites converge, conflict, and transmute."[1]

Code/Coding

Coding is the process by which humans give the computer a detailed set of instructions to be executed in sequential order. The set of careful step-by-step instructions, which frequently references other nested instructions, creates a system of rules, or the code.

Code Work

The social practices that surround the *code worlds*, taking into account the production, circulation, and reception of narratives, artifacts, and subjectivities that arise within hacking collectives. Code work is the labor my research participants engage in as they create and think with the *ethno-stack*.

---

1. Keating, 2009: 10.

Code Worlds

The space and time coders inhabit when they become immersed in computer programming, or *coding.* To explore the code worlds anthropologically means to consider how coding is inherently a social practice embedded in specific cultural and political contexts, influenced by power structures that mold the computing infrastructures themselves.

Documentation

Text accompanying *code* that helps someone else understand how it works or why a developer made particular implementation decisions. Might be embedded in the code itself or in documents accompanying the code.

Ethno-Stack

The *ethno* in ethno-stack points first to the definition of "ethno-" as a particular culture or people, this notion of difference signaling the different stacks that can emerge from stack theorizing. The *ethno* also refers to the *ethnographic* approach that can lead us to think in this more expansive way about computing and the code worlds. The ethno- and ethnographic in the ethno-stack thus work together to ground the stack, to ask how it might be inhabited, contested, accommodated, resisted, multiplied, situated, or bent.

Across the book's chapters, I offer four basic layers that can help us think with the ethno-stack: the personal, the interpersonal, the sociopolitical, and the sociotechnical.

While each implementation of the ethno-stack can call for different research questions across its layers, the following questions can help us think with the generic ethno-stack:

How and why do people choose to enter the stack?

What are the different layers that make up this particular stack?

How do people use metaphors from one layer of the stack to solve problems in another layer of the stack, or another domain of life?

What type of work does it take to move between the layers of the stack?

How does difference in social position (along dimensions of race, gender, class, disability, sexuality) influence the way users navigate the stack?

How do other(ed) communities propose we reconfigure the stack?

What might alternative stacks look like?

Full-Stack Developer

A software developer who shows interest and mastery in navigating all layers of the application stack. A common way to describe a full-stack developer, for example, is as someone who can write code for both the back-end of a project (e.g., databases, architecture, servers) and the front-end of a project (e.g., graphical user interfaces, web applications, mobile clients).

Hackathon

The hackathon is the ritual event for hacker-entrepreneurs in the 21st century. It takes on very different modes of participation depending on the group of organizers and sponsors, but the basic idea of the form is that an interdisciplinary group of (mostly) young people meet and network with other hackers or entrepreneurs over the course of a weekend. In a span of 48–72 hours, participants are expected to meet partners, develop technological or technocratic solutions to a problem related to an organizing theme (e.g., healthcare, transportation, immigration), and pitch their startup to investor-judges. The pitch must convey why the startup is an innovative project, what problem it is resolving, and many times, that it is scalable and economically viable in the market.

Hacker

I use the term "hacker" to refer to someone who loves to program computers in the spirit of playfulness and exploration. While my designation of "hacker" to any particular research participant corresponds to the multitude of definitions of *hacking*, I usually only call someone a hacker if this is how they self-identify, usually because they possess the technical skills to immerse themselves in the *code worlds* and program computers.

Hacker-Entrepreneur

This hybrid term points to the way many research participants often shift between these identities, or in some cases see no difference between them at all. The term points to this fluidity between identities but also to the ways techno-entrepreneurial Silicon Valley–esque cultures have come to dominate hacking cultures in Mexico and Latin America more broadly.

Hacker Ethics

Developed at the Mexico City hacker school[2], the 10 principles guiding hackers:

<1>   Give before you get

<2>   No pedir permiso (Don't ask for permission)

---

2. Morato, 2015a, 2015b.

<3>   Hacer > Hablar (Doing > Talking)

<4>   No existen excusas (No Excuses)

<5>   Resolver problemas (Solve problems)

<6>   Sigue tu curiosidad (Follow your curiosity)

<6.2> Fracasar = Crecer (Failing = Growing)

<7>   Conoce tus herramientas y comunidades (Know your tools/communities)

<8>   Siempre aprender (Always be learning)

<9>   Involucrarse (Get involved)

<10>   Divertirse en el proceso (Have fun)

Hacker School

Founded in Mexico City in 2014 by hackers who wanted to "live the hackathon every day," this bootcamp style program is built on the idea that Mexico's social and political ills might be resolved if Mexico had more hackers. With an itinerant bootcamp-style program, expert coders train hackers-in-the-making to master their programming skills, work collaboratively on creative projects, and cultivate the hacker ethos made explicit by the ten principles of their *hacker ethic.*

Hackerspace

Usually refers to a physical space where people hack, of more permanent nature than the hackathon, but not always. For the purposes of this book, hackerspace encompasses the hackathon as well, i.e., the hackathon also counts as a hackerspace.

Hacking

Definitions of hacking vary according to which origin story or typology we subscribe to. Hacking can include some aspects of repurposing technology for means other than for what it was intended; playful tinkering (which usually involves computation); technical competency; or remixing old and new media infrastructures with grassroots organizing. Across its various expressions, hacking can be seen as a site where craft and craftiness converge.[3] A hack generally connotes a clever technical solution arrived at through non-obvious means. The hacker identity, then, might be taken up by (or assigned to) "hacktivists, free software developers, hacker-entrepreneurs, hack-driven leakers and journalists, criminal extorters of bitcoin, or information security researchers in search of a safer internet."[4]

3. Coleman, 2017a: 161.
4. Coleman and Kelty, 2017.

Loop/Iteration

A fundamental *coding* concept, a loop tells a computer program to perform a set of actions until the conditions defined for the loop are met. Each pass through the loop is called an iteration. A computer programmer who need to run the same lines of *code* many times can save time by using a loop.

Loose Coupling

A computing term that refers to a robust way to write code where data structures (or other components) can use other components in an inter-connected system without needing to know the full details of their implementation. In this way, each component becomes more autonomous and can be used for different purposes by different components; elements become "coupled" and depend on each other with very little (or no) direct knowledge of each other.

Minimal Viable Product (MVP)

Especially in more commercially-focused hackathons, a software program (or "product") with just enough features to attract the first users in order to be able to validate an early idea in the product development cycle. Subsequent iterations of the product incorporate feedback from these early working versions.

Open-Source Software

Software that is distributed with its source code, the code behind computer programs that users rarely ever see, which coders use to control how their programs behave. This makes the *code* available for use, modification, and distribution by other coders. An accompanying license typically allows programmers to modify the software and control how the new versions of the software can be distributed. Open-source software is usually developed in a decentralized and collaborative way, relying on peer review and community production.

Pivot

A buzzword in the tech startup world, it refers to being flexible with your technological project (many times an *app*) and changing it quickly to something that sticks with users, and, in most cases, with investors. In other words, a "pivot" more closely aligns your product with market dynamics.

Prototype

As part of the software development process, a rudimentary software program (or product) that is by definition incomplete, representative of the final program. After enough *iterations*, the program is expected to reach its final form.

The Stack

The stack refers to the interrelated and interdependent layers of hardware components and software protocols that make the high-level computations and programs possible. More abstractly, to move from the bottom of the stack (e.g., machine code) to the top of the stack (e.g., programming languages and systems) means to traverse the corresponding circuits, microchips, and computer code that can be part of each layer of abstraction that makes up the system. The fundamental idea is that one can navigate the stack by building up layers of abstraction from lower-level components. Across different layers of the stack, each configuration of elements becomes a component to be used by other components. The corresponding internal implementation of each element is abstracted away and largely irrelevant to the other components that use it.

Techno-Borderlands

My notion of the techno-Borderlands pays homage to, and also plays with, Anzaldúa's concept of Borderlands to examine the political economy of tech and code work as it relates to US–México politics and the corresponding subjectivities of difference found apart from yet deeply influenced by the US/México border. In this zone, US/Mexican hackers deploy the language of coding to navigate boundaries of nation, race, ethnicity, class, and gender—boundaries, often hierarchical, that coding promises to reconfigure, but that also remain essential to the transnational economy of tech and resurgent at every step.

User

Especially in the context of the *hackathon*, the imagined human who will ultimately use the *app*. Typically distinguished by the coders themselves as someone who does not have *coding* know-how, i.e., the coders design and implement the app; the users use the app.

Cast of Code Workers

ASU = Arizona State University
FOIA = Freedom of Information Act
CA = California
CDMX = Ciudad de México
ESCOM = Escuela Superior de Cómputo
ITESM = Instituto Tecnológico y de Estudios Superiores de Monterrey
MIT = Massachusetts Institute of Technology
UV = Universidad Veracruzana
UNAM = Universidad Nacional Autónoma de México
IPN = Instituto Politécnico Nacional
SF = San Francisco
SV = Silicon Valley
US = United States

CODE_WORKER =

[Name]
> fieldsites;
> age; educational or professional background;
> relevant geographic or national origin;
> other pertinent biographic details;
> misc lifestyle/cultural preferences (if available);
// appears in [ ].[ ]

[ALICIA]

> CDMX women's hackathon;
> early 20s; recent graduate of visual arts from La Esmeralda art school;
> moved from Guanajuato to study in CDMX;

> thinks there should have been more men at the women's hackathon;
> ultimate frisbee athlete;
// appears in [4].[2]

[ARMIOS]

> hackathons in both US and Mexico; Migrahack Denver;
> mid-30s; studied computer science and engineering at UC Davis;
> born in SF Bay Area to Mexican migrants; double nationality, US
    and Mexican;
> tech contract work allows him geographic flexibility to do remote work
> new job with major tech company in CA, looking to settle down
    permanently with family in Bay Area;
// appears in [3].[3], [4].[3], [5].[2], [5].[3], [5].[4]

[AZUKITA]

> iLab Xalapa;
> early 20s; studied chemical engineering at UV;
> recruited to iLab from Minatitlan, Veracruz;
> diligently works on Re-Active, an app aimed at reducing chronic
    body pain;
> carries her bunny rabbit "Petri" with her most places she goes;
// appears in [1].[1]

[CESAR]

> Migrahack Los Angeles;
> late 30s; journalist by training;
> US Latinx, born and raised in City of Industry, CA;
> newbie to the code worlds;
> diehard Dallas Cowboys fan;
// appears in [5].[2], [5].[3]

[CHAVITA]

> many hackathons in Mexico; hacker school;
> early 20s; studied computer science at UNAM;
> from near Torres de Satelite in Naucalpan, Estado de México;

> well-regarded sensei at hacker school in CDMX; repeatedly wins
    hackathon competitions; co-founder of competitive coding club
    at UNAM;
> claims the best pozole is served at a secret hole in the wall in Plaza
    Garibaldi;
// appears in [1].[0], [1].[3], [1].[4], [1].[5], [2].[0], [2].[1]

[CINDY]

> Migrahack CDMX, Tucson;
> late 20s; studied Education, Social Justice and Human Rights
    at ASU;
> US Latinx, born and raised in Phoenix, AZ;
> works for nonprofit which documents human rights violations
    within US immigration centers;
> loves Frida Kahlo (of course);
// appears in [5].[2]

[COFI]

> many hackathons in Mexico; hacker school;
> early 20s; studied computer systems engineering at ESCOM at IPN;
> from Ecatepec de Morelos, Estado de México;
> dad/"jefe" is an auto mechanic;
> earned this name because of the amount of coffee he consumes;
    girlfriend named his laptop "Mildred;" he would have preferred
    "Daisy";
// appears in [2].[0], [2].[1], [2].[2], [2].[4], [3].[1], [3].[2], [3].[3]

[DANIELA RIVERO]

> appearance at women's hackathon;
> late 20s; studied visual arts at UV;
> born in Atizapan de Zaragoza, Estado de Mexico; lives in Xalapa;
> inspired by world-class prehispanic pieces in Xalapa's Museo de
    Antropología;
> family believes grandfather, career electrician before he died,
    communicates with world of the living via refrigerator;
// appears before [1].[0], [2].[0], [3].[0], [4].[0], [5].[0], [6].[0]

[EL PATO]

> iLab Xalapa; various hackathons and tech entrepreneur events in Mexico;

> mysterious age; studied Business Administration and Information Technology at ITESM;

> from Córdoba, Veracruz; previously worked for large corporate companies;

> received certificate in Innovation and Technology from MIT Entrepreneurship program, presents himself as "MIT graduate";

> earned his nickname because of his overuse of phrase, "Yo escopeta, tú pato" (Me shotgun, you duck);

// appears in [1].[2], [1].[3]

["ESTEBAN," ALSO HÉCTOR BELTRÁN (AUTHOR)]

> all fieldsites;

> mid 20s to mid 30s; ambivalent anthropologist, unrelenting code worker;

> Mexican/American; from San Fernando Valley, CA;

> MIT computer science undergrad, now MIT anthropology professor;

> favorite short story is "El ahogado más hermodo del mundo," by García Marquéz;

// appears in [2].[2], [2].[4], [6].[3]

[ESTEFY]

> iLab Xalapa;

> late 20s; studied Visual Arts at UV;

> recruited to iLab from Boca del Rio, Veracruz;

> zealously worked on SugarNut, a vibrator that anonymous users controlled over wifi;

> avid PokémonGo player;

// appears in [3].[2]

[FERNANDO]

> Migrahack Chicago;
> mid 30s; editor for Spanish-language newspaper; experienced bilingual investigative reporter;
> US Latinx; Chicago born and raised;
> experience with manipulating datasets, submitting FOIA requests;
> watching *Seinfeld* puts him to sleep;
// appears in [5].[2]

[HIRO]

> hackathons in both US and Mexico; SF dating scene;
> late 20s; studied electrical engineering and product design; builds hardware prototypes for early-stage startups;
> born in Mexico to Guatemalan father and Japanese mother; grew up in Mexico speaking Japanese and Spanish; fluent in English and sign language;
> well-liked entrepreneurial lead in MIT coding bootcamp at UNAM;
> leaves his framed photograph in Airbnb rentals;
// appears in [3].[2], [3].[4]

[ÍO]

> women's hackathon;
> early 20s; mechatronics student at UNAM and graphic design enthusiast;
> moved to study in CDMX from Delicias, Chihuahua;
> started social media accounts to circulate information about legal abortions in MX;
> developed her love for engineering by hacking her Furby babies;
// appears in [4].[2], [4].[3]

[JAVO]

> "Mexico Day" in Sunnyvale; various hackathons and tech entrepreneur events in US and Mexico;

> mid to late 20s; International Relations and MBA at Universidad de Monterrey;
> raised in Monterrey; grandmother from El Paso, TX;
> mother is a teacher, father is a medical doctor; previously teacher at both an under-resourced public high school and one of the top private high schools in Mexico;
> advice to other entrepreneurs is to read *The Lean Startup* early; guiding principles developed to deal with ethical situation are "Don't lie and don't steal."

// appears in [0].[0], [0].[3], [6].[0], [6].[1], [6].[2], [6].[3], [6].[4], [6].[5]

[JESSEE]

> SV tech startup scene;
> late 30s; studied Symbolic Systems at Stanford, attended Berkeley Law School;
> US Latinx;
> adamant home schooler;
> thinks using socks with sandals shows bad taste;

// appears in [5].[0]

[KIKE]

> hacker school co-founder; most hackathons in MX; SV tech startup scene;
> late 20s; studied computer science and engineering at ESCOM at IPN;
> proud CDMX native;
> previously co-founded Androidtitlán, community of Android developers focused on sourcing, translating, and sharing resources in Spanish;
> loves eating at Casa Toño not necessarily because of the food, but because of their efficient operations, and especially the teamwork between waiters;

// appears in [1].[3]

[LEO]

> most hackathons in Mexico; MIT-UNAM coding bootcamp;
> early 20s; studied electrical engineering at UNAM;
> from Chicoloapan, Estado de México;
> winner of several hackathons in Mexico, including some that earned him trips to Google and Facebook headquarters; all his savings go to making trips to visit friends who work for companies in SF and SV;
> favorite book is *Hackers and Painters*;
// appears in [1].[2], [1].[3], [1].[4], [2].[0], [2].[2], [3].[3]

[LOTAR]

> iLab Xalapa;
> early 30s; studied Design and Visual Communication at UV, self-taught programmer;
> moved to study in Xalapa from Oaxaca;
> speaks Zapoteco, English, Spanish;
> proud owner of a 1930s Chandler & Price movable type press;
// appears in [1].[1], [3].[1]

[LUIS]

> various hackathons in Mexico; hacker school; MIT-UNAM coding bootcamp;
> early 30s; studied applied mathematics at UNAM;
> from "La Loma" Estado de Mexico;
> top scorer in programming assessment test for bootcamp;
> proud dog father of adopted canine characters such as "El gitano," "El siete mugres," "Chiclan," "La Pikuni," and "La Turca";
// appears in [3].[1]

[MARIANA]

> MIT-UNAM coding bootcamp; various hackathons in Mexico; women's hackathon;
> early 20s; studies computer systems engineering at ESCOM at IPN;

> proudly born in Santa Julia, "barrio mágico" in CDMX best
known for "El Tigre de Santa Julia," legendary bandit during el
Porfiriato;
> 1 of only 4 women participants in UNAM-MIT bootcamp; highly
skilled coder;
// appears in [4].[2], [4].[3]

[MEMO]

> various hackathons in Mexico; MIT-UNAM coding bootcamp;
> early 20s; studied computer science and engineering at UNAM;
> from Nezahualcóyotl, Estado de México;
> recruited to Seattle to work for popular media streaming
company;
> frequently posts pictures of different local brews he has tried;
committed faux pas when he showed up in suit to first MIT
bootcamp session;
// appears in [1].[4], [6].[2]
// *not to be confused with other Memo, character in Sleep Dealer*
discussed in coda

[MONICA]

> Migrahack Los Angeles;
> early 40s; studied Ethnic Studies and later Journalism at
UC Berkeley;
> Chicana from San Jose, CA;
> career in publishing media specializing in cities with large
Latino populations;
> studying to be wine connoisseur; claims you can find best
wines at Trader Joe's;
// appears in [5].[2]

[RODO]

> iLab Xalapa; MIT coding bootcamp Xalapa;
> late 20s; studied computational technologies, UV, master's in
artificial intelligence, UV;
> born and raised in Xalapa;

> spends free time romhacking, creating modifications of video games
(especially translations) and releasing them on his blog;
> father would walk him daily to front of university and kiss him to
embarrass him;
// appears in [3].[2]

[SAIPH]

> women's hackathon; various hackathons in US and Mexico;
> mid 20s; studied computer engineering at UNAM, computer science
PhD at UC Santa Barbara;
> organizing lead of women's hackathon and relentless advocate for
higher education;
> both parents professors; father head of a bio-robotics laboratory
at UNAM;
// appears in [4].[0], [4].[1], [4].[2]

[T.C. OR TECNOCHICA]

> women's hackathon;
> early 20s; mechatronics student at UNAM;
> moved to study in CDMX from Cosoleacaque, Veracruz;
> travels frequently between CDMX and Veracruz;
> hates fandango; claims her best ideas come during long
bus rides home;
// appears in [3].[1], [4].[2]

Featured Figurillas

Prehispanic figures making appearances in exvotos that open up Chapters [1]–[6]:

["PERSONAJE CON ANTEOJERAS"]

Cultura de Centro de Veracruz. Clásico Tardío 600–900 n.e. El Zapotal, Veracruz. 41 × 26 × 48 cm. Museo de Antropología de Xalapa, Veracruz.

// *appears in Exvotos [1], "Python el dios de los 0s y 1s"*

["FIGURA SONRIENTE"]

Cultura de Centro de Veracruz. Clásico Tardío 600–900 n.e. El Zapotal, Veracruz. 33 × 19.5 × 10 cm.

// *appears in Exvotos [2] = "Ángel Guardián del Hacking"*

["EL COLUMPIO"]

Cultura de Centro de Veracruz. 300–900 n.e. Procedencia desconocida. 13.5 × 22 × 16 cm. Museo de Antropología de Xalapa, Veracruz.

// *appears on bed in Exvotos [3] = "Cristo del curso Zoom de las masculinidades diversas"*

["FIGURILLA FEMENINA"]

Cultura Preclásico del Altiplano. Preclásico Medio. Tlapacoya, estado de México. 13.6 × 8.6 cm. Museo Nacional de Antropología.

// *appears in window in Exvotos [3] = "Cristo del curso Zoom de las masculinidades diversas"*

["CIHUATETEO"]

Cultura de Centro de Veracruz. Clásico Tardío 600–900 n.e. El Zapotal, Veracruz. 52 × 33 cm. Museo de Antropología de Xalapa, Veracruz.

// *appears on far left in Exvotos [4] = "Santo Niño de Arduino amigo mío"*

["MÚSICO"]

Cultura Preclásico del Altiplano. Preclásico Medio. Tlatilco, estado de México. 11.5 × 6.2 cm. Museo Nacional de Antropología.

// *appears on floor with laptop in Exvotos [4] = "Santo Niño de Arduino amigo mío"*

["FIGURILLA FEMENINA"]

Cultura de Centro de Veracruz. Clásico. El Faisán, Veracruz. 34.5 cm 29.1 cm. Museo Nacional de Antropología.

// *appears hovering on cloud in Exvotos [4] = "Santo Niño de Arduino amigo mío"*

["MICTLANTECUHTLI"]

Azteca-mexica. Posclásico Tardío (ca. 1500). 176 × 80 cm. Museo del Templo Mayor, ciudad de México.

// *appears in Exvotos [5] = "Virgencita del Agile programming"*

["EL SEÑOR DE LAS LIMAS"]

Olmeca. 900–400 a.n.e. Las Limas, Jesús Carranza, Veracruz. 55 × 43.5 × 23 cm. Museo de Antropología de Xalapa, Veracruz.

// *appears in Exvotos [6] = "Santo Señor del Pivoteo"*

REFERENCES

Abbate, Janet. *Recoding Gender.* Cambridge, MA: MIT Press, 2012.

Abbate, Janet, and Stephanie Dick. "Introduction: Thinking with Computers." In *Abstractions and Embodiments: New Histories of Computing and Society*, ed. Jane Abbate and Stephanie Dick, 1–19. Baltimore, MD: Johns Hopkins University Press, 2022.

Adler Lomnitz, Larissa, Claudio Lomintz Adler, and Ilya Adler. "El fondo de la forma: La campaña presidencial del PRI en 1988." *Nueva Antropología* 11, no. 38 (1990): 45–82.

Alarcón, Rafael. "Recruitment Processes among Foreign-Born Engineers and Scientists in Silicon Valley." *American Behavioral Scientist* 42, no. 9 (1999): 1382–97.

Althusser, Louis. "Ideology and Ideological State Apparatuses (Notes toward an Investigation)." In *Lenin and Philosophy and Other Essays,* trans. Ben Brewster, 142–7, 166–76. New York NY: Monthly Review Press, 1971.

Ames, Morgan G. *The Charisma Machine: The Life, Death, and Legacy of One Laptop per Child.* Cambridge, MA: MIT Press, 2019.

Ames, Morgan G., Silvia Lindtner, Shaowen Bardzell, Jeffrey Bardzell, Lilly Nguyen, Syed Ishtiaque Ahmed, Nusrat Jahan, Steven J. Jackson, and Paul Dourish. "Making or Making Do? Challenging the Mythologies of Making and Hacking." *Journal of Peer Production* 12 (2018): 1–21.

Amit, Vered, and Noel Dyck. "Pursuing Respectable Adulthood: Social Reproduction in Time of Uncertainty." In *Young Men in Uncertain Times,* ed. Vered Amit and Noel Dyck, 1–31. New York, NY: Berghahn, 2012.

Amrute, Sareeta. *Encoding Race, Encoding Class: Indian IT Workers in Berlin.* Durham, NC: Duke University Press, 2016.

Amrute, Sareeta. "What Would a Techno-Ethics Look Like?" *Platypus: The Committee on the Anthropology of Science, Technology, and Computing,* January 6, 2018. http://blog.castac.org /2018/01/techno-ethics/ (accessed February 6, 2023).

Amrute, Sareeta, and Luis Felipe Murillo. "Introduction: Computing in/from the South." *Catalyst: Feminism, Theory, and Technoscience* 6, no. 2 (2020): 1–23.

Anand, Nikhil, Akhil Gupta, and Hannah Appel. *The Promise of Infrastructure.* Durham, NC: Duke University Press, 2018.

Anzaldúa, Gloria. *Borderlands La Frontera: The New Mestiza.* San Francisco, CA: Aunt Lute, 1987.

Anzaldúa, Gloria. *Interviews/Entrevistas.* Ed. AnaLouise Keating. New York: Routledge, 2000.

Aparicio, Frances. "Latinidad/es." In *Keywords for Latina/o Studies,* eds. Deborah R. Vargas, Nancy Raquel Mirabal, and Lawrence La Fountain-Stokes, 113–17. New York, NY: New York University Press, 2017.

Aparicio, Frances. *Negotiating Latinidad: Intralatina/o Lives in Chicago.* Urbana, IL: University of Illinois Press, 2019.

Archambault, Julie Soleil. *Mobile Secrets: Youth, Intimacy, and the Politics of Pretense in Mozambique.* Chicago, IL: University of Chicago Press, 2017.

Avle, Seyram, Silvia Lindtner, and Kaiton Williams. "How Methods Make Designers." *Proceedings of SIGCHI Conference on Human Factors in Computing Systems* (CHI '17, 2017): 472–83.

Barbrook, Richard, and Andy Cameron. "The Californian Ideology." *Science as Culture* 6, no. 1 (2009): 44–72.

Barney, Darin, Gabriella Coleman, Christine Ross, Jonathan Sterne, and Tamar Tembeck. *The Participatory Condition in the Digital Age*. Minneapolis: University of Minnesota Press, 2016.

Bateson, Gregory. *Steps to an Ecology of Mind*. Chicago, IL: University of Chicago Press, 1972.

Beltrán, Cristina. *The Trouble with Unity: Latino Politics and the Creation of Identity*. New York: Oxford University Press, 2010.

Beltrán, Héctor. "'The Right to *El Mall*'. Book review of: Dávila, Arlene. 2016. *El Mall: The Spatial and Class Politics of Shopping Malls in Latin America*." *Anthropology Now* 9, no. 2 (2017a): 121–5.

Beltrán, Héctor. "Proficiency: Provocation." Correspondences, *Cultural Anthropology* website, June 12, 2017b. https://culanth.org/fieldsights/proficiency-provocation (accessed February 15, 2023).

Beltrán, Héctor. "Latina/os and Tech: Toward a Holistic Approach for Diversifying Silicon Valley." *Latina/os and Tech Initiative Policy Brief*. Center for Latino Policy Research, UC Berkeley, 2017c.

Beltrán, Héctor. "Code Work: Thinking with the System in México." *American Anthropologist* 122, no.3 (2020a): 487–500.

Beltrán, Héctor. "The First Latina Hackathon: Recoding Infrastructures from México." *Catalyst: Feminism, Theory, and Technoscience* 6, no.2 (2020b): 1–30.

Beltrán, Héctor. "Cybraceros: The Promise and Perils of Border Hacking." *Hack_Curio*, September 7, 2021. https://hackcur.io/cybraceros-the-promise-and-perils-of-border-hacking/ (accessed February 15, 2023).

Beltrán, Héctor. "Hacking, Computing Expertise, and Difference." *Just Tech*, Social Science Research Council, March 1, 2022. https://doi.org/10.35650/JT.3029.d.2022.

Berardi, Franco "Bifo." *The Soul at Work: From Alienation to Autonomy*. South Pasadena, CA: Semiotext(e), 2009.

Blackwell, Maylei. *Chicana Power! Contested Histories of Feminism in the Chicano Movement*. Austin, TX: University of Texas Press, 2011.

Boellstorff, Tom. "Zuckerberg and the Anthropologist: Facebook, Culture, Digital Futures." *Culture Digitally*, February 27, 2017. https://culturedigitally.org/2017/02/zuckerberg-and-the-anthropologist-facebook-culture-digital-futures/ (accessed February 10, 2023).

Boellstorff, Tom. "The Opportunity to Contribute: Disability and the Digital Entrepreneur." *Information, Communication, & Society* 22, no. 4 (2019): 474–90.

Boltanski, Luc, and Eva Chiapello. *The New Spirit of Capitalism*, trans. Gregory Elliott. London: Verso, 2007.

Booth, William. "Mexico Is Now a Top Producer of Engineers, but Where Are Jobs?" *Washington Post*, October 28, 2012. https://www.washingtonpost.com/world/the_americas/mexico-is-now-a-top-producer-of-engineers-but-where-are-jobs/2012/10/28/902db93a-1e47-11e2-8817-41b9a7aaabc7_story.html (accessed February 1, 2023).

Bratton, Benjamin H. *The Stack: On Software and Sovereignty*. Cambridge, MA: MIT Press, 2016.

Briggs, Charles L. *Learning How to Ask: A Sociolinguistic Appraisal of the Role of the Interview in Social Science Research*. Cambridge: Cambridge University Press, 1986.

Briggs, Charles L. "Anthropology, Interviewing, and Communicability in Contemporary Society." *Current Anthropology* 48, no. 4 (2007): 551–80.

Cacho, Lisa. *Social Death: Racialized Rightlessness and the Criminalization of the Unprotected*. New York: New York University Press, 2012.

Candelario, Ginetta E.B. "Transnationalism." In *Keywords for Latina/o Studies*, eds. Deborah R. Vargas, Nancy Raquel Mirabal, and Lawrence La Fountain-Stokes, 236–8. New York, NY: New York University Press, 2017.

Casillas, Dolores Inés. *Sounds of Belonging: U.S. Spanish-Language Radio and Public Advocacy.* New York, NY: New York University Press, 2014.

Castellanos, M. Bianet. "Becoming Chingón/a: A Gendered and Racialized Critique of the Global Economy." In *Strange Affinities: The Gender and Sexual Politics of Comparative Racialization*, ed. Grace Kyungwon Hong and Roderick A. Ferguson, 270–92. Durham, NC: Duke University Press, 2011.

Chappell, Ben. "'Take a Little Trip with Me': Lowriding and the Poetics of Scale." In *Technicolor: Race, Technology, and Everyday Life*, ed. Alondra Nelson and Thuy Linh N. Tu with Alicia Headlamp Hines, 100–20. New York, NY: New York University Press, 2001.

Chavez, Leo. *Shadowed Lives: Undocumented Immigrants in American Society.* Fort Worth, TX: Harcourt Brace Jovanovich, 1992.

Chavez, Leo R. *Covering Immigration: Popular Images and the Politics of the Nation.* Berkeley, CA: University of California Press 2001.

Chavez, Leo R. *The Latino Threat: Constructing Immigrants, Citizens, and the Nation.* Stanford, CA: Stanford University Press, 2008.

Chavez, Leo R. *Anchor Babies and the Challenge of Birthright Citizenship.* Stanford, CA: Stanford University Press, 2017.

Chun, Wendy Hui Kyong. *Programmed Visions: Software and Memory.* Cambridge, MA: MIT Press, 2013.

Chun, Wendy Hui Kyong, Winnie Soon, Noah Wardrip-Fruin, and Jichen Zhu. "Software Studies, Revisited." *Computational Culture* (May 9, 2022, as pre-print). http://computationalculture.net/software-studies-revisited/ (accessed February 7, 2023).

CodersLink. *Tech Salaries in Mexico: 2019 Report.* https://coderslink.com/company/tech-salaries-report-2019/ (accessed February 15, 2023).

Coleman, E. Gabriella. *Coding Freedom: The Ethics and Aesthetics of Hacking.* Princeton, NJ: Princeton University Press, 2013.

Coleman, E. Gabriella. *Hacker, Hoaxer, Whistleblower, Spy: The Many Faces of Anonymous.* New York: Verso, 2014.

Coleman, E. Gabriella. "Hacker." In *Digital Keywords: A Vocabulary of Information Society and Culture*, ed. Benjamin Peters, 158–72. Princeton, NJ: Princeton University Press, 2017a.

Coleman, E. Gabriella. "From Internet Farming to Weapons of the Geek." *Current Anthropology* 58, no. S15 (2017b): S91–S102.

Coleman, E. Gabriella, and Alex Golub. "Hacker Practice." *Anthropological Theory* 8, no. 3 (2008): 255–77.

Coleman, E. Gabriella, and Christopher Kelty. "Preface: Hacks, Leaks, and Breaches." LIMN no. 8, (2017). https://limn.it/issues/hacks-leaks-and-breaches/ (accessed February 15, 2023).

Collier, Stephen J. "Topologies of Power: Foucault's Analysis of Political Government beyond 'Governmentality'." *Theory Culture Society* 26, no. 6 (2009): 78–108.

Combahee River Collective. "A Black Feminist Statement." In *Capitalist Patriarchy and the Case for Socialist Feminism*, ed. Zillah R. Eisenstein, 362–72. New York, NY: Monthly Review Press, 1979.

Corsín Jiménez, Alberto. "The Prototype: More than Many and Less than One." *Journal of Cultural Economy* 7, no. 4 (2014): 381–98.

Costanza-Chock, Sasha. *Design Justice: Community-Led Practices to Build the Worlds We Need.* Cambridge, MA: MIT Press, 2020.

Crenshaw, Kimberlé. "Demarginalizing the Intersection of Race and Sex: A Black Feminist Critique of Antidiscrimination Doctrine, Feminist Theory and Antiracist Politics." *University of Chicago Legal Forum* 1 (1989): 139–68.

Crooks, Roderic N. "Times Thirty: Access, Maintenance, and Justice." *Science, Technology, and Human Values* 44, no. 1 (2018): 118–42.

Crossa, Veronica. "Resisting the Entrepreneurial City: Street Vendors' Struggle in Mexico City's Historic Center." *International Journal of Urban and Regional Research* 33, no. 1 (2009): 45–63.

da Costa Marques, Ivan. "Cloning Computers: From Rights of Possession to Rights of Creation." *Science as Culture* 14 (2005): 139–60.

Dávila, Arlene. *Latinos, Inc.: The Marketing and Making of a People.* Berkeley, CA: University of California Press, 2001.

Dávila, Arlene. *Latino Spin: Public Image and the Whitewashing of Race.* New York, NY: New York University Press, 2008.

Dávila, Arlene. *El Mall: The Spatial and Class Politics of Shopping Malls in Latin America.* Oakland, CA: University of California Press, 2016.

de Certeau, Michel. *The Practice of Everyday Life.* Berkeley: University of California Press, 1984.

De Genova, Nicholas. *Working the Boundaries: Race, Space, and "Illegality" in Mexican Chicago.* Durham, NC: Duke University Press, 2005.

De Genova, Nicholas, and Ana Y. Ramos-Zayas. *Latino Crossings: Mexicans, Puerto Ricans, and the Politics of Race and Citizenship.* New York, NY: Routledge, 2003.

DeHart, Monica. *Ethnic Entrepreneurs: Identity and Development Politics in Latin America.* Stanford, CA: Stanford University Press, 2010.

Dhaliwal, Ranjodh Singh. "The Cyber-Homunculus: On Race and Labor in Plans for Computation." *Configurations* 30 (2022): 377–409.

D'Ignazio, Catherine, Alexis Hope, Alexandra Metral, Willow Brugh, David Raymond, Becky Michelson, Tal Achituv, and Ethan Zuckerman. "Towards a Feminist Hackathon: The 'Make the Breast Pump Not Suck!' Hackathon." *Journal of Peer Production,* no. 8 (2016). http://peerproduction.net/issues/issue-8-feminism-and-unhacking-2 (accessed January 30, 2023).

Dunbar-Hester, Christina. *Low Power to the People: Pirates, Protest, and Politics in FM Radio Activism.* Cambridge, MA: MIT Press, 2014.

Dunbar-Hester, Christina. "Paradoxes of Participation." In *The Participatory Condition in the Digital Age,* ed. Darin Barney, Gabriella Coleman, Christine Ross, Jonathan Sterne, and Tamar Tembeck, 79–99. Minneapolis: University of Minnesota Press, 2016.

Dunbar-Hester, Christina. *Hacking Diversity: The Politics of Inclusion in Open Technology Cultures.* Princeton, NJ: Princeton University Press, 2020.

The Economist Staff. "Mexico and the United States: The Rise of Mexico." *Economist,* November 24, 2012. http://www.economist.com/news/leaders/21567081-america-needs-look-again-its-increasingly-important-neighbour-rise-mexico/ (accessed February 1, 2023).

English-Lueck, J.A. *Cultures@SiliconValley.* Stanford, CA: Stanford University Press, 2002.

Ensmenger, Nathan. "Making Programming Masculine." In *Gender Codes: Why Women are Leaving Computing,* ed. Thomas Misa, 115–41. Hoboken, NJ: John Wiley & Sons, 2010.

Escobar Latapí, Agustin, and Laura Patricia Pedraza Espinoza. "Clases medias en México: Transformación social, sujetos multiples." In *Clases Medias en América Latina: Retrospectiva y cambios recientes,* ed. Rolando Franco, Martín Hopenhayn, and Arturo León, 355–408. Mexico City: Siglo XXI editores, 2010.

Félix, Adrián. *Specters of Belonging: The Political Life Cycle of Mexican Migrants.* New York, NY: Oxford University Press, 2019.

Fish, Adam, Luis F.R. Murillo, Lilly Nguyen, Aaron Panofksy, and Christopher Kelty. "Birds of the Internet: A Field Guide to Understanding Action, Organization, and the Governance of Participation." *Journal of Cultural Economy* 4, no. 2 (2011): 157–87.

Fleming, Rachel C. "'Work Is Just for Timepass.' Gendered Inversions of Productive and Non-productive Time for Women Technology Workers in Bangalore." *Voices* 13 (2018): 88–98.

Fortun, Kim. "Ethnography in Late Industrialism." *Cultural Anthropology* 27, no. 3 (2012): 446–64.

Fortun, Kim. "From Latour to Late Industrialism." *HAU: Journal of Ethnographic Theory* 4, no. 1 (2014): 309–29.

Fouché, Rayvon. *Black Inventors in the Age of Segregation*. Baltimore, MD: Johns Hopkins University Press, 2003.

Fought, Carmen. *Language and Ethnicity: Key Topics in Sociolinguistics*. Cambridge: Cambridge University Press, 2006.

Freeman, Carla. *Entrepreneurial Selves: Neoliberal Respectability and the Making of a Caribbean Middle Class*. Durham, NC: Duke University Press, 2014.

Friedman, Thomas L. "How Mexico Got Back in the Game." *New York Times,* February 23, 2013. http://www.nytimes.com/2013/02/24/opinion/sunday/friedman-how-mexico-got-back-in -the-game.html?_r=0 (accessed February 1, 2023).

Fuller, Matthew, ed. *Software Studies: A Lexicon*. Cambridge, MA: MIT Press, 2008.

Gaboulry, Jacob. "A Queer History of Computing." *Rhizome,* February 19, 2013. https://rhizome .org/editorial/2013/feb/19/queer-computing-1/ (accessed February 1, 2023).

Gal, Susan. "Contradictions of Standard Language in Europe: Implications for the Study of Practices and Publics." *Social Anthropology* 13, no. 2 (2006): 163–81.

García, Lorena. *Respect Yourself, Protect Yourself: Latina Girls and Sexual Identity*. New York, NY: New York University Press, 2012.

García Canclini, Néstor. *Culturas híbridas: Estrategias para entrar y salir de la modernidad*. Mexico DF: Editorial Grijalbo, 2009 [1990].

García Canclini, Néstor. "Introducción: De la cultura postindustrial a las estrategias de los jóvenes." In *Jóvenes, Culturas Urbanas y Redes Digitales: Prácticas emergentes en las Artes, Las Editoriales y la Música*, ed. by Néstor García Canclini, Francisco Cruces, and Maritza Urteaga Castro Pozo, 3–24. Madrid: Fundación Telefónica, 2012.

García Canclini, Néstor, and Francisco Cruces. "Conversación a modo de prólogo." In *Jóvenes, Culturas Urbanas y Redes Digitales: Prácticas emergentes en las Artes, Las Editoriales y la Música*, ed. Néstor García Canclini, Francisco Cruces, and Maritza Urteaga Castro Pozo, i–x. Madrid: Fundación Telefónica, 2012.

Gauntlet, David. *Making Is Connecting*. Cambridge: Polity, 2011.

Gelber, Steven M. "Do-It-Yourself: Constructing, Repairing and Maintaining Domestic Masculinity." *American Quarterly* 49, no. 1 (1997): 66–112.

Gershon, Ilana. "Seeing Like a System: Luhmann for Anthropologists." *Anthropological Theory* 5, no. 2 (2005): 99–116.

Gershon, Ilana. *Down and Out in the New Economy: How People Find (or Don't Find) Work Today*. Chicago, IL: The University of Chicago Press, 2017.

Gershon, Ilana. "Employing the CEO of Me, Inc.: US Corporate Hiring in a Neoliberal age." *American Ethnologist* 45, no. 2 (2018): 173–85.

Gil, Liliana. "A Fablab at the Periphery: Decentering Innovation from São Paulo." *American Anthropologist* 124, no. 4 (2022): 721–33.

Giles, Martin. "Here's How Hackers Could Cause Chaos in This Year's US Midterm Elections." *MIT Technology Review,* April 18, 2018. https://www.technologyreview.com/s/610774/heres -how-hackers-could-cause-chaos-in-this-years-us-midterm-election/ (accessed February 13, 2023).

Ginsburg, Faye D., Lila Abu-Lughod, and Brian Larkin. "Introduction." In *Media Worlds: Anthropology on New Terrain*, ed. Faye D. Ginsburg, Lila Abu-Lughod, and Brian Larkin, 1–36. Berkeley, CA: University of California Press, 2002.

Ginsburg, Faye, and Rayna Rapp. "Disability/Anthropology: Rethinking the Parameters of the Human." *Current Anthropology* 61, no. S21 (2020): S4–S15.

Gomberg-Muñoz, Ruth. "Willing to Work: Agency and Vulnerability in an Undocumented Immigrant Network." *American Anthropologist* 112, no. 2 (2010): 295–307.

Gómez-Peña, Guillermo. "The Virtual Barrio @ The Other Frontier: (or The Chicano Interneta)." In *Technicolor: Race, Technology, and Everyday Life*, eds. Alondra Nelson and Thuy Linh N. Tu with Alicia Headlamp Hines, 191–98. New York, NY: New York University Press, 2001.

Graham, Elyse. "Punk Culture and the Rise of the Hacker Ethic." In *Abstractions and Embodiments: New Histories of Computing and Society*, ed. Jane Abbate and Stephanie Dick, 380–98. Baltimore, MD: Johns Hopkins University Press, 2022.

Gregg, Melissa. "FCJ-186 Hack for Good: Speculative Labor, App Development and the Burden of Austerity." *Fibreculture Journal* 25 (2015): 183–201.

Gregg, Melissa. *Counterproductive: Time Management in the Knowledge Economy*. Durham, NC: Duke University Press, 2018.

Gregory, Sam. "The Participatory Panopticon and Human Rights: WITNESS's Experience Supporting Video Advocacy and Future Possibilities." In *Sensible Politics: The Visual Culture of Nongovernmental Activism*, ed. Meg McLagan and Yates McKee, 517–49. New York, NY: Zone, 2012.

Guidotti-Hernández, Nicole M. "Borderlands." In *Keywords for Latina/o Studies*, ed. Deborah R. Vargas, Nancy Raquel Mirabal, and Lawrence La Fountain-Stokes, 21–24. New York, NY: New York University Press, 2017.

Gupta, Akhil, and James Ferguson. "Beyond 'Culture': Space, Identity, and the Politics of Difference." *Cultural Anthropology* 7 (1992): 6–23.

Gupta, Akhil, and James Ferguson. "Discipline and Practice: 'The Field' as Site, Method, and Location in Anthropology." In *Anthropological Locations: Boundaries and Grounds of a Field Science*, ed. Akhil Gupta and James Ferguson, 1–46. Berkeley, CA: University of California Press, 1997.

Haraway, Donna. "A Cyborg Manifesto: Science, Technology, and Socialist-Feminism in the Late Twentieth Century." In *Simians, Cyborgs, and Women: The Reinvention of Nature*, Donna Haraway, 149–82. New York, NY: Routledge, 1991.

Haring, Kristen. *Ham Radio's Technical Culture*. Cambridge, MA: MIT Press, 2006.

Harrell, Fox. *Phantasmal Media: An Approach to Imagination, Computation, and Expression*. Cambridge, MA: MIT Press, 2013.

Heiman, Rachel, Carla Freeman, and Mark Liechty. *The Global Middle Classes: Theorizing Through Ethnography*. Santa Fe, NM: School for Advanced Research Press, 2012.

Helmreich, Stefan. *Silicon Second Nature: Culturing Artificial Life in a Digital World*. Berkeley, CA: University of California Press, 1998.

Helmreich, Stefan. *A Book of Waves*. Durham, NC: Duke University Press, 2023.

Himanen, Pekka. *The Hacker Ethic and the Spirit of the Information Age*. New York, NY: Random House, 2001.

Ho, Karen. *Liquidated: An Ethnography of Wall Street*. Durham, NC: Duke University Press, 2009.

Hoffman, Lisa M. *Patriotic Professionalism in Urban China: Fostering Talent*. Philadelphia, PA: Temple University Press, 2010.

hooks, bell. *All about Love*. New York, NY: Harper Collins, 2001.

Howell, Jayne. "Getting Out to Get Ahead? Perspectives on Schooling and Social and Geographic Mobility in Southern Mexico." *Journal of Latin American and Caribbean Anthropology* 23, no. 2 (2017): 301–19.

Iglesias-Prieto, Norma. "Maquiladoras." In *Keywords for Latina/o Studies,* ed. Deborah R. Vargas, Nancy Raquel Mirabal, and Lawrence La Fountain-Stokes, 125–29. New York, NY: New York University Press, 2017.

Instituto Mexicano de la Juventud. *Encuesta Nacional de Juventud 2010.* México: IMJUVE, 2010. http://politicasdejuventud.celaju.net/wp-content/uploads/2014/05/Encuesta-Juv-2010.pdf (accessed February 1, 2023).

Irani, Lilly. "Hackathons and the Making of Entrepreneurial Citizenship." *Science, Technology, & Human Values* 40, no. 5 (2015): 799–824.

Irani, Lilly. *Chasing Innovation: Making Entrepreneurial Citizens in Modern India.* Princeton, NJ: Princeton University Press, 2019.

Irani, Lilly. "Hackathons: Labor, Politics, and the Organization of Public Passions." In *Lives of Data: Essays on Computational Cultures from India*, ed. Sandeep Mertia, 68–78. Amsterdam: Institute of Network Cultures, 2020.

Irvine, Judith, and Susan Gal. "Language Ideology and Linguistic Differentiation." In *Linguistic Anthropology: A Reader*, ed. Alessandro Duranti, 402–34. Oxford: Blackwell, 2001.

Jackson, Liz. "We Are the Original Lifehackers." *New York Times*, May 30, 2018. https://www.nytimes.com/2018/05/30/opinion/disability-design-lifehacks.html (accessed February 7, 2023).

Jeffrey, Craig. "Timepass: Youth, Class, and Time among Unemployed Men in India." *American Ethnologist* 37, no. 3 (2010): 465–81.

Johnson, Amy, and Graham Jones. "Language, the Internet, and Digital Communication." In *The International Encyclopedia of Linguistic Anthropology*, ed. J. M. Stanlaw, 1–13. Malden, MA: Wiley-Blackwell, 2021.

Jones, Graham M., Beth Semel, and Audrey Le. "'There's no rules. It's a hackathon.': Negotiating Commitment in a Context of Volatile Sociality." *Journal of Linguistic Anthropology* 25, no. 3 (2015): 322–45.

Joo, Rachel Miyung. *Transnational Sport: Gender, Media, and Global Korea.* Durham, NC: Duke University Press, 2012.

Jordan, Tim. "A Genealogy of Hacking." *Convergence: The International Journal of Research into New Media Technologies* 23, no. 5 (2017): 528–44.

Joyce, Patrick. *The Rule of Freedom: Liberalism and the Modern City.* London: Verso, 2003.

Keating, AnaLouise. "Introduction: Shifting Worlds, una Entrada." In *Entre Mundos/Among Worlds: New Perspectives on Gloria Anzaldúa*, ed. AnaLouise Keating, 1–12. New York, NY: Palgrave Macmillan, 2005.

Keating, AnaLouise. "Introduction: Reading Gloria Anzaldúa, Reading Ourselves . . . Complex Intimacies, Intricate Connections." In *The Gloria Anzaldúa Reader,* ed. AnaLouise Keating, 1–15. Durham, NC: Duke University Press, 2009.

Kelty, Christopher M. *Two Bits: The Cultural Significance of Free Software.* Durham, NC: Duke University Press, 2008.

Kelty, Christopher M. *The Participant: A Century of Participation in Four Stories.* Chicago, IL: University of Chicago Press, 2019.

Kothari, Ashish, Ariel Sallah, Arturo Escobar, Federico Demaria, and Alberto Acosta. *Pluriverse: A Post-Development Dictionary.* New Delhi: Tulika, 2019.

Krishnan, Sneha. "Doing Nothing: Gender, Respectability, and Playing with Time." *Voices* 13, no. 1 (2018): 62–73.

Larkin, Brian. *Signal and Noise: Media, Infrastructure, and Urban Culture in Nigeria.* Durham, NC: Duke University Press, 2008.

Leal Martínez, Alejandra. "'You Cannot Be Here': The Urban Poor and the Specter of the Indian in Neoliberal Mexico City." *Journal of Latin American and Caribbean Anthropology* 21, no. 3 (2016): 539–59.

Levy, Steven. *Hackers: Heroes of the Computer Revolution.* Sebastopol, CA: O'Reilly, 2010 [1984].

Lewis, Jason. "Preparations for a Haunting: Notes toward an Indigenous Future Imaginary." In *The Participatory Condition in the Digital Age,* ed. Darin Barney, Gabriella Coleman, Christine Ross, Jonathan Sterne, and Tamar Tembeck, 229–49. Minneapolis, MN: University of Minnesota Press, 2016.

Li, Tania Murray. *The Will to Improve: Governmentality, Development, and the Practice of Politics.* Durham, NC: Duke University Press, 2007.

Light, Jennifer. "When Computers Were Women." *Technology & Culture* 40 (1999): 455–83.

Limón, José. *Dancing with the Devil: Society and Cultural Poetics in Mexican-American South Texas.* Madison, WI: University of Wisconsin Press, 1994.

Lindtner, Silvia. *Prototype Nation: China and the Contested Promise of Innovation.* Princeton, NJ: Princeton University Press, 2020.

LiPuma, Edward, and Benjamin Lee. *Financial Derivatives and the Globalization of Risk.* Durham, NC: Duke University Press, 2004.

Liu, Alan. "Towards a Diversity Stack: Digital Humanities and Diversity as Technical Problem." *PMLA* 135, no. 1 (2020): 130–51.

Lombana-Bermudez, Andres, Sandra Cortesi, Christian Fieseler, Urs Gasser, Alexa Hasse, Gemma Newlands, and Sarah Wu. "Youth and the Digital Economy: Exploring Youth Practices, Motivations, Skills, Pathways, and Value Creation." *Berkman Klein Center Research Publication* no. 2020-4 (2020). https://papers.ssrn.com/sol3/papers.cfm?abstract_id=3622572 (accessed January 30, 2023).

Lomnitz, Claudio. *Deep Mexico, Silent Mexico: An Anthropology of Nationalism.* Minneapolis, MN: University of Minnesota Press, 2001.

Magaña, Rafael Maurice. *Cartographies of Youth Resistance: Hip-Hop, Punk, and Urban Autonomy in Mexico.* Oakland, CA: University of California Press, 2020.

Marez, Curtis. *Farm Worker Futurism: Speculative Technologies of Resistance.* Minneapolis, MN: University of Minnesota Press, 2016.

Martin, Emily. "Flexible Survivors." *Anthropology News* 40, no. 6 (1999): 5–7.

Marwick, Alice E. *Status Update: Celebrity, Publicity, and Branding in the Social Media Age.* New Haven, CT: Yale University Press, 2013.

Maxigas, Peter. "Hacklabs and Hackerspaces: Tracing Two Genealogies." *Journal of Peer Production* 2 (2012). http://peerproduction.net/issues/issue-2/peer-reviewed-papers/hacklabs-and-hackerspaces (accessed January 30, 2023).

Mazzarella, William. *Shoveling Smoke: Advertising and Globalization in Contemporary India.* Durham, NC: Duke University Press, 2005.

McIntosh, Lukas and Caroline D. Hardin. "Do Hackathon Projects Change the World? An Empirical Analysis of GitHub Repositories." *Proceedings of ACM Technical Symposium on Computer Science Education* (SIGCSE '21, 2021): 879–85.

McLagan, Meg. "Imagining Impact: Documentary Film and the Production of Political Effects." In *Sensible Politics: The Visual Culture of Nongovernmental Activism,* ed. Meg McLagan and Yates McKee, 305–21. New York, NY: Zone, 2012.

Mead, Margaret. "Cybernetics of Cybernetics." In *Purposive Systems: Proceedings of the Fifth Annual Symposium of the American Society of Cybernetics,* ed. Heinz von Foerster, John D. White, Larry J, Peterson, and John K. Russell, 1–11. New York: Spartan Books, 1968.

Medina, Eden, Ivan da Costa Marques, and Christina Holmes. "Introduction: Beyond Imported Magic." In *Beyond Imported Magic: Essays on Science, Technology, and Society in Latin America,* ed. Eden Medina, Ivan da Costa Marques, and Christina Holmes, 1–24. Cambridge, MA: MIT Press, 2014.

Mendoza-Denton, Norma. *Homegirls: Language and Cultural Practice among Latina Youth Gangs*
  Hoboken, NJ: Blackwell, 2008.

Meneses, Maria Paula. "Tastes, Aromas, and Knowledges: Challenges to a Dominant Epistemology."
  In *Knowledges Born in the Struggle: Constructing the Epistemologies of the Global South*, ed.
  Boaventura de Sousa Santos and Maria Paula Meneses, 162–80. New York, NY: Routledge,
  2020.

Mitchell, Timothy. *Rule of Experts: Egypt, Techno-politics, Modernity*. Berkeley, CA: University
  of California Press, 2002.

Mitter, Swasti. "Information Technology and Working Women's Demands." In *Women Encounter*
  Technology: Changing Patterns of Employment in the Third World, ed. Swasti Mitter and Sheila
  Rowbotham, 19–43. New York, NY: Routledge, 1995.

Miyazaki, Hirokazu. "The Temporalities of the Market." *American Anthropologist* 105, no. 2 (2003):
  255–65.

Miyazaki, Hiro. "Economy of Dreams: Hope in Global Capitalism and Its Critiques." *Cultural*
  Anthropology 21, no. 2 (2006): 147–72.

Montoya, Rosario. "Women's Sexuality, Knowledge, and Agency in Rural Nicaragua." In *Gender's*
  Place: Feminist Anthropologies of Latin America, ed. Rosario Montoya, Lessie Jo Frazier, and
  Janise Hurtig, 65–88. New York, NY: Palgrave Macmillan, 2002.

Moore, Henrietta L. "Concept-Metaphors and Pre-Theoretical Commitment in Anthropology."
  Anthropological Theory 4, no. 1 (2004): 71–88.

Mora, Cristina G. *Making Hispanics: How Activists, Bureaucrats, and Media Constructed a New*
  American. Chicago, IL: The University of Chicago Press, 2014.

Mora, Mariana. "Ayotzinapa and the Criminalization of Racialized Poverty in La Montaña, Guer-
  rero, Mexico." *PoLAR: Political and Legal Anthropology Review* 40, no. 1 (2017): 67–85.

Moraga, Cherríe, and Gloria E. Anzaldúa. *This Bridge Called My Back: Writings by Radical Women*
  of Color. Watertown, MA: Persephone Press, 1981.

Morato, Manuel. "El Decálogo de la Cultura Hacker." *Medium,* January 31, 2015a. https://medium
  .com/@ememorato/el-decalogo-de-la-cultura-hacker-95a6a45b5d1b (accessed February 24,
  2023).

Morato, Manuel. "The Ten Commandments of Hacker Culture." *Medium,* February 1, 2015b. https://
  medium.com/@ememorato/the-ten-commandments-of-hacker-culture-4e183d570eb6
  (accessed February 24, 2023).

Mukherjee, Sanjukta. "Producing the Knowledge Professional: Gendered Geographies of Alien-
  ation in India's New High-Tech Workplace." In *In an Outpost of the Global Information Econ-
  omy: Work and Workers in India's Outsourcing Industry*, ed. Carol Upadhya and AR Vasavi
  50–75. New Delhi: Routledge, 2008.

Nakamura, Lisa. "Indigenous Circuits: Navajo Women and the Racialization of Early Electronic
  Manufacture." *American Quarterly* 66, no. 4 (2014): 919–41.

Nguyen, Lilly U. "Infrastructural Action in Vietnam: Inverting the Techno-Politics of Hacking in
  the Global South." *New Media & Society* 18, no. 4 (2016): 637–52.

Nieto, Mercedes Pedrero. "Genero, trabajo doméstico y extradomestico en México. Una esti-
  mación del valor económico del trabajo doméstico." *Estudios Demográficos y Urbanos* 19,
  no. 2 (2004): 413–46.

Olguín, B.V. "Contrapuntal Cyborgs?: The Ideological Limits and Revolutionary Potential of
  Latin@ Science Fiction." *In Altermundos: Latin@ Speculative Literature, Film, and Popular*
  Culture, ed. Cathryn Merla-Watson and B.V. Olguín, 128–44. Los Angeles, CA: UCLA Chicano
  Studies Research Center Press, 2017.

Ong, Aihwa. "Cultural Citizenship as Subject-Making: Immigrants Negotiate Racial Cultural
  Boundaries in the United States." *Current Anthropology* 37, no. 5 (1996): 737–62.

Ong, Aihwa. *Flexible Citizenship: The Cultural Logics of Transnationality*. Durham, NC: Duke University Press, 1999.

Ong, Aihwa. *Neoliberalism as Exception: Mutations in Citizenship and Sovereignty*. Durham, NC: Duke University Press, 2006.

Oyĕwùmí, Oyèrónké. *The Invention of Women: Making an African Sense of Western Gender Discourses*. Minneapolis, MN: University of Minnesota Press, 1997.

Padilla, Tanalís. "Espionage and Education: Reporting on Student Protest in Mexico's Normales Rurales, 1960–1980." *Journal of Iberian and Latin American Research* 19, no. 1 (2013): 20–29.

Padilla, Tanalís. *Unintended Lessons of Revolution: Student Teachers and Political Radicalism in Twentieth-Century Mexico*. Durham, NC: Duke University Press, 2022.

Parés, Gustavo. "In the Race for Tech Talent, the US Should Look to Mexico." *TechCrunch,* May 20, 2012. https://techcrunch.com/2021/05/20/in-the-race-for-tech-talent-the-us-should-look-to-mexico/ (accessed February 13, 2023).

Paz, Octavio. *The Labyrinth of Solitude*, trans. Lysander Kemp, Yara Milos, and Rachel Phillips-Belash. New York, NY: Grove Press, 1985.

Pérez, Gina M. *Citizen, Student, Soldier: Latina/o Youth, JROTC, and the American Dream*. New York, NY: New York University Press, 2015.

Pertierra, Anna Cristina. "Practicing Tranquilidad: Domestic Technologies and Comfortable Homes in Southeastern Mexico." *Journal of Latin American and Caribbean Anthropology* 20, no. 3 (2015): 415–32.

Peterson, Nicole D. "'We Are Daughters of the Sea': Strategies, Gender, and Empowerment in a Mexican Women's Cooperative." *Journal of Latin American and Caribbean Anthropology* 19, no. 1 (2014): 148–67.

Philip, Kavita, Lilly Irani, and Paul Dourish. "Postcolonial Computing: A Tactical Survey." *Science, Technology, and Human Values* 37, no. 1 (2012): 3–29.

Poggiali, Lisa. "Seeing (from) Digital Peripheries: Technology and Transparency in Kenya's Silicon Savannah." *Cultural Anthropology* 31, no. 3 (2016): 387–411.

Popescu, Adam. "Is Mexico the Next Silicon Valley? Tech Boom Takes Root in Silicon Valley." *The Washington Post,* May 14, 2016. https://www.washingtonpost.com/business/is-mexico-the-next-silicon-valley-tech-boom-takes-root-in-guadalajara/2016/05/13/61249f36-072e-11e6-bdcb-0133da18418d_story.html?utm_term=.f88e2313d08c (accessed February 15, 2023).

Portillo, David Morison, and Rihan Yeh. *Border Vueltas/Looping Fronterizo*. San Diego, CA: Taller California, 2021.

Quenardel, P.J. "Infographic: Worldwide Hackathon Figures in 2016." *Bemyapp,* January 4, 2017. https://agency.bemyapp.com/insights/infographics-hackathon-figures-in-2016.html (accessed January 28, 2023).

Radhakrishnan, Smitha. 2008. "Examining the 'Global' Indian Middle Class: Gender and Culture in the Silicon Valley/Bangalore Circuit." *Journal of Intercultural Studies* 29, no. 1 (2008): 7–20.

Ramírez, Catherine S. "Alternative Cartographies: Third Woman and the Respatialization of the Borderlands." *Midwestern Miscellany* 30 (2004a): 47–62.

Ramírez, Catherine S. "Deus Ex Machina: Tradition, Technology, and the Chicanafuturist Art of Marion C. Martinez." *Aztlán* 29, no. 2 (2004b): 55–92.

Ramírez, Catherine S. *The Woman in the Zoot Suit: Gender, Nationalism, and the Cultural Politics of Memory*. Durham, NC: Duke University Press, 2009.

Ramos-Zayas, Ana. *Street Therapists: Race, Affect, and Neoliberal Personhood in Latino Newark*. Chicago, IL: University of Chicago Press, 2012.

Rea, Stephen C. "Calibrating Play: Sociotemporality in South Korean Digital Gaming Culture." *American Anthropologist* 120, no. 3 (2018): 500–11.

Reagle, Joseph M., Jr. *Hacking Life: Systematized Living and Its Discontents.* Cambridge, MA: MIT Press, 2019.

Reguillo, Rosanna. "La condición juvenil en el México contemporáneo: Biografías, incertidumbres y lugares." In *Los jóvenes en México*, ed. Rosanna Reguillo, 395–430. México: FCE/ Conaculta, 2010.

Ricuarte Quijano, Paola. "Jóvenes y cultural digital: Abordes críticos desde América Latina." *Chasqui: Revista Latinoamericana de Comunicación* 137 (2018): 13–28.

Rios, Victor M. *Punished: Policing the Lives of Black and Latino Boys.* New York, NY: New York University Press, 2011.

Rivera, Alex. *Why Cybraceros?* http://alexrivera.com. Alex Rivera website, 1997. http://alexrivera .com/project/why-cybraceros/ (accessed February 12, 2023)

Rivera, Alex, dir. *Sleep Dealer.* Maya Entertainment, 2008.

Rosa, Fernanda. "Code Ethnography and the Materiality of Power in Internet Governance." *Qualitative Sociology* 45 (2022): 433–55.

Rosa, Jonathan. "Standardization, Racialization, Languagelessness: Raciolinguistic Ideologies across Communicative Contexts." *Journal of Linguistic Anthropopology* 26, no. 2 (2016): 162–83.

Rosa, Jonathan. *Looking Like a Language, Sounding Like a Race: Raciolinguistic Ideologies and the Learning of Latinidad.* Oxford: Oxford University Press, 2019.

Rosa, Jonathan, and Nelson Flores. "Unsettling Race and Language: Toward Raciolinguistic Perspective." *Language and Society* 46 (2017): 621–47.

Rosaldo, Renato. *Culture and Truth: The Remaking of Social Analysis.* Boston, MA: Beacon 1989.

Rosales, F. Arturo. *Chicano! The History of the Mexican American Civil Rights Movement.* Houston, TX: Arte Público, 1997.

Rosas, Gilberto. "The Thickening Borderlands: Bastard Mestiz@s, 'Illegal' Possiblities, and Globalizing Migrant Life." In *Critical Ethnic Studies: A Reader*, ed. Nada Elia, David M. Hernández, Jodi Kim, Shana L. Redmond, Dylan Rodriguez, and Sarita Echavez See, 344–59. Durham, NC: Duke University Press, 2016.

Rose, Nikolas. *Powers of Freedom: Reframing Political Thought.* Cambridge: Cambridge University Press, 1999.

Salzinger, Leslie. "Re-Marking Men: Masculinity as a Terrain of the Neoliberal Economy." *Critical Historical Studies* 3, no. 1 (2016): 1–25.

Sandoval, Chela. *Methodology of the Oppressed.* Minneapolis, MN: University of Minnesota Press, 2000.

Santa Anna, Otto. *Brown Tide Rising: Metaphors of Latinos in Contemporary Public Discourse.* Austin, TX: University of Texas Press, 2002.

Santos, Boaventura de Sousa. *Epistemologies of the South: Justice Against Epistemicide.* New York, NY: Routledge, 2014.

Santos, Boaventura de Sousa, and Maria Paula Meneses. *Knowledges Born in the Struggle: Constructing the Epistemologies of the Global South.* New York, NY: Routledge, 2020.

Saxenian, AnnaLee. *Regional Advantage: Culture and Competition in Silicon Valley and Route 128.* Cambridge, MA: Harvard University Press, 1996.

Saxenian, AnnaLee. *The New Argonauts: Regional Advantage in a Global Economy.* Cambridge, MA: Harvard University Press, 2006.

Schrock, Andrew R. "Civic Hacking as Data Activism and Advocacy: A History from Publicity to Open Government Data." *New Media & Society* 18, no. 4 (2016): 581–99.

Seaver, Nick. "Algorithms as Culture: Some Tactics for Ethnography of Algorithmic Systems." *Big Data & Society*, July–December 2017, 1–12.

Seaver, Nick. "What Should an Anthropology of Algorithms Do?" *Cultural Anthropology* 33, no. 3 (2018): 375–85.

Seaver, Nick. "Captivating Algorithms: Recommender Systems as Traps." *Journal of Material Culture* 24, no. 4 (2019): 421–36.

Shankar, Shalini. *Desi Land: Teen Culture, Class, and Success in Silicon Valley*. Durham, NC: Duke University Press, 2008.

Sims, Christo. *Disruptive Fixation: School Reform and the Pitfalls of Techno-Idealism*. Princeton, NJ: Princeton University Press, 2017.

Söderberg, Johan, and Alessandro Delfanti. "Hacking Hacked! The Life Cycles of Digital Innovation." *Science, Technology, and Human Values* 40, no. 5 (2015): 793–98.

Sterne, Jonathan. "Bourdieu, Technique, and Technology." *Cultural Studies* 17, no. 3 (2003): 367–89.

Strathern, Marilyn. *Reproducing the Future: Anthropology, Kinship, and the New Reproductive Technologies*. Manchester: Manchester University Press, 1992.

Suchman, Lucy, Randall Trigg, and Jeanette Blomberg. "Working Artefacts: Ethnomethods of the Prototype." *British Journal of Sociology* 53, no. 2 (2002): 163–79.

Talavera, Victor, Núñez-Mchiri, Guillermina Gina, Heyman, Josiah. "Deportation in the U.S.-Mexico Borderlands: Anticipation, Experience, and Memory." In *The Deportation Regime: Sovereignty, Space, and the Freedom of Movement*, ed. Nicholas De Genova and Nathalie Peutz. 166–95. Durham, NC: Duke University Press, 2010.

Terranova, Tiziana. "Red Stack Attack! Algorithms, Capital and the Automation of the Common." In *#Accelerate: The Accelerationist Reader*, ed. Robin Mackay and Armen Avanessian, 379–99. Falmouth: Urbanomic, 2014.

Tigau, Camelia. *Riesgos de la Fuga de Cerebros: Construcción Mediática, Posturas Gubernamentales, y Expectativas de los Migrantes*. Ciudad Universitaria: Universidad Nacional Autónoma de México, 2013.

Turkle, Sherry. *The Second Self: Computers and the Human Spirit*. New York, NY: Simon & Schuster, 1984.

Turner, Fred. *From Counterculture to Cyberculture: Stewart Brand, the Whole Earth Network, and the Rise of Digital Utopianism*. Chicago, IL: University of Chicago Press, 2006.

Turner, Fred. "Prototype." In *Digital Keywords: A Vocabulary of Information Society & Culture*, ed. Benjamin Peters, 256–68. Princeton, NJ: Princeton University Press, 2016.

Urciolli, Bonnie. "Skills and Selves in the New Workplace." *American Ethnologist* 35, no. 2 (2008): 211–28.

Uribe, Verónica del Águila. "Culturas de prototipado de caretas durante la pandemia: Comunicación y economozacíion de la participación cívica en México." *Conexión* 10, no. 16 (2021): 175–93.

Urteaga Castro Pozo, Maritza. "De jóvenes, contemporáneos, *Trendys*, emprendedores y empresarios culturales." In *Jóvenes, Culturas Urbanas y Redes Digitales: Prácticas emergentes en las Artes, Las Editoriales y la Música*, ed. Néstor García Canclini, Francisco Cruces, and Maritza Urteaga Castro Pozo, 25–44. Madrid: Fundación Telefónica, 2012.

Valdez, Mónica. "Jóvenes y datos. Panorama de la desigualdad." In *Juventudes, culturas, identidades y tribus juveniles en el México con-temporáneo*, ed. Maritza Urteaga, *Suplemento Diario de Campo*, no. 56 (2009): 37–9.

Valenzuela, José Manuel. *Juvenicidio: Ayotzinapa y las Vidas Precarias en América Latina y España*. Barcelona: Nuevos Emprendimientos Editoriales, 2015.

Villegas, Paulina. "Disenchanted Youth May Tip Mexican Election to López Obrador." *New York Times*, June 25, 2018. https://www.nytimes.com/2018/06/25/world/americas/mexico-election-youth.html (accessed February 1, 2023).

Wark, McKenzie. *A Hacker Manifesto*. Cambridge, MA: Harvard University Press, 2004.

Weizenbaum, Joseph. *Computer Power and Human Reason: From Judgement to Calculation*. San Francisco, CA: W.H. Freeman, 1976.

Wortham, Stanton, Katherine Mortimer, and Elaine Allard. "Mexicans as Model Minorities in the New Latino Diaspora." *Anthropology and Education Quarterly* 40, no. 4 (2009): 388–404.

Yeh, Rihan. "Deslices del 'mestizo' en la frontera norte." In *Nación y alteridad: Mestizos, indígenas y extranjeros en el proceso de formación nacional*, ed. Daniela Gleizer and Paula López Caballero 405–36. México: Universidad Autónoma Metropolitana Unidad Cuajimalpa, 2015.

Yeh, Rihan. *Passing: Two Publics in a Mexican Border City*. Chicago, IL: University of Chicago Press, 2018.

Yergeau, Melanie. "Disability Hacktivism." In "Hacking the Classroom: Eight Perspectives," ed. Mary Hocks and Jentery Sayers, *Computers and Composition Online* 2014. http://cconline journal.org/hacking/#yergeau (Accessed February 7, 2023).

Yost, Jeffrey R. *Making IT Work: A History of the Computer Services Industry*. Cambridge, MA: MIT Press, 2017.

Zaloom, Caitlin. "The Productive Like of Risk." *Cultural Anthropology* 19, no.3 (2004): 365–391.

Zandbergen, Dorien. "Silicon Valley New Age: The Co-Constitution of the Digital and the Sacred." In *Religions and Modernity: Relocating the Sacred to the Self and the Digital,* ed. Stef Aupers and Dick Houtman, 161–85. Leiden: Brill, 2010.

Zentella, Ana Celia. "Latina/o Languages and Identities." In *Latinos: Remaking America,* ed. Manuel Suarez-Orozco and Mariela Páez, 321–28. Berkeley, CA: University of California Press, 2002.

Zilberg, Elana. *Space of Detention: The Making of a Transnational Gang Crisis between Los Angeles and San Salvador*. Durham, NC: Duke University Press, 2011.

Zukin, Sharon, and Max Papadantonakis. "Hackathons as Co-optation Ritual: Socializing Workers and Institutionalizing Innovation in the 'New' Economy." In *Precarious Work: Causes, Characteristics, and Consequences. Research in the Sociology of Work*, vol. 31, ed. Arne L. Kalleberg and Steven P. Vallas, 157–81. Bingley UK: Emerald, 2017.

INDEX

A NOTE ON THE TYPE

This book has been composed in Adobe Text and Gotham. Adobe Text, designed by Robert Slimbach for Adobe, bridges the gap between fifteenth- and sixteenth-century calligraphic and eighteenth-century Modern styles. Gotham, inspired by New York street signs, was designed by Tobias Frere-Jones for Hoefler & Co.

GPSR Authorized Representative: Easy Access System Europe - Mustamäe tee 50, 10621 Tallinn, Estonia, gpsr.requests@easproject.com

www.ingramcontent.com/pod-product-compliance
Ingram Content Group UK Ltd.
Pitfield, Milton Keynes, MK11 3LW, UK
UKHW042250300325
456820UK00002B/23